Practical Project Management
for Engineers

For a listing of recent titles in the
Artech House Effective Project Management Library,
turn to the back of this book.

Practical Project Management for Engineers

Nehal Patel

ARTECH HOUSE

BOSTON | LONDON

artechhouse.com

Library of Congress Cataloging-in-Publication Data
A catalog record for this book is available from the U.S. Library of Congress.

British Library Cataloguing in Publication Data
A catalog record for this book is available from the British Library.

ISBN-13: 978-1-63081-585-1

Cover design by John Gomes

© 2019 Artech House
685 Canton Street
Norwood, MA 02062

10 9 8 7 6 5 4 3 2 1

To my beautiful Yogi Nivas and Roulo Family!!
James and Faye Roulo, Mahendrabhai and Hemuben Patel,
Kamleshbhai and Pinkyben Patel, Jayantibhai and Apuben Patel.

Contents

Acknowledgments   *xiii*

Introduction   *xv*
> References   *xxi*

CHAPTER 1

Communications Management   1

Know the Audience   2
Effective Communication Media   3
Email Use   5
> Email Distribution   6

Phones and Voice Mail   9
Effective Meetings   10
> Meeting Preparation   10
> No Meeting Wednesday   10
> Use WebEx   10

Direct, Informal Communication   11
> Allow Back Channels   11
> Heads Up   12
> Be Accessible   12

The Importance of Templates   12
> Trip Reports   12
> Weekly Status: Four Corner Charts (Quad Chart)   12
> 80-20 Report   14
> Template Evolution   14

Conclusion   15
> References   15

CHAPTER 2

Scope Management   17

Plan Project Management   18
Plan Scope Management   21
Collect Project Requirements   24

Document Project Scope   26
Create WBS   27
Validate Scope   29
Control Scope   29
Summary   31
    References   32

CHAPTER 3

Schedule Management   33

Define Activities   37
    Work Breakdown Structure   37
    Work Package Planning   38
    Indirect Efforts   39
    Phase Gate Reviews   40
    Level of Detail   41
    Rolling Wave Plan   42
    Invoicing Strategy   43
Sequence Activities   43
    Activity Dependencies   43
    Network Diagrams   44
Estimate Activity Duration   45
    Duration and Elapsed Time   47
    Start and Finish Dates   47
    Activity Duration Rollup   48
    Resource Loading   49
    Duration Estimating Techniques   49
    Reserve Analysis   52
Estimate Activity Resources   52
    Resource Requirements   53
Develop Schedule: Analyze and Baseline   53
    Schedule Analysis   53
    Schedule Baselines   55
Use the Schedule   56
    Tracking Progress   56
    Assessing Performance   57
    Analyzing Performance   57
    Critical Path Analysis   58
    Corrective Actions   59
Summary   61
    References   61

CHAPTER 4

Requirements Management   63

Identify User Needs   66
Collect Requirements   68

Elaborate Requirements 71
Express Requirements 74
Analyze Requirements 76
Requirements Verification and Validation 82
Summary 84
References 84

CHAPTER 5

Risk Management 85

Plan Risk Management 88
Identify Risks 91
Risk Identification Categories 95
Risk Identification Statement 97
Analyze Risks 98
Handle Risks 102
Monitor Risks 104
Opportunity Management 105
Summary 106
References 107

CHAPTER 6

Project Resource Management 109

Organizational Environment 109
Plan Resource Management 114
Acquire Project Resources 117
Develop Project Team 120
Manage the Project Team 123
Summary 125
References 125

CHAPTER 7

Vendor Management 127

Vendor Selection 130
Vendor Product/Service Acceptance 140
Monitor Vendor Performance 144
Summary 147
References 148

CHAPTER 8

Cost Management 149

Organize the Project 153
Develop Project Schedule and Budget 156
Define Measurement Methods and Baseline the Project 160
Analyze Project Performance 163

Summary   168
    References   169

CHAPTER 9

Configuration Management   171

Plan Configuration Management   172
Identify Configuration Items   175
Manage Configuration Changes   175
Maintain Configuration Item Status   178
Evaluate Configuration Items and Changes   179
Release New Configuration Items   179
Summary   180
    References   180

CHAPTER 10

Quality Management   181

Plan Quality Management   184
Quality Assurance   187
Quality Control   190
Balanced Quality Management   193
Summary   194
    References   195

CHAPTER 11

Tales from the Trenches   197

Communication Management   197
    The Value of Informal Communications   197
Scope Management   198
    Three Perspectives of Scope Management from a Single Program   198
    The Mars Viking Lander Project   200
Schedule Management   202
    Key Takeaways   203
Requirements Management   203
    AMRAAM: Write the Specification to Win the Contract!   204
    Communicate Clearly with the Contractor/Subcontractor   204
    View Requirements through the Eyes of the Customer   205
Risk Management   206
Resource Management   207
    Project: Transition Hospital Operations to a New State-of-the-Art Facility   207
Vendor Management   209
Cost Management   211
    Planning   211
    Planning at the Right Level   212
    Culture of Accountability   212

Monitoring Cost and Schedule Management   212
The Importance of Cost Awareness   212
Quality Management   214
Configuration Management   216
The AWG-9 Radar for the F-14 Aircraft: A Persistent Test Failure   216

About the Author   219

Index   221

Acknowledgments

I owe special thanks to my friends for providing real project management stories retold throughout the book in *Tales from the Trenches*. Without their contribution the book would not be nearly as practical.

- Eric J. Roulo, aerospace structures consultant, president of Roulo Consulting Inc.
- Richard "DJ" Johnson, senior vice president at Booz Allen Hamilton
- Dr. Fred Brown, graduate director at Loyola Marymount University (retired)
- Dr. Don Macleod, chief executive officer at Applied Motion Products
- John (Jack) Conrad, director of program management at Raytheon (retired)
- Colonel Janet Grondin, United States Air Force (retired)
- Dr. Tejash Gandhi, chief operating officer at Baystate Medical Center
- Rich Fitzgerald, chief operating officer at SMTC Corporation
- David Kubera, senior program manager at Defense Aerospace Corporation.
- Larry Johnson, president of LR Johnson Associates LLC
- John D. Lillard, senior fellow, founder of Capability Assurance Institute

I am also profoundly thankful for my parents. Without their continuous sacrifice, I would not be in a position to make it this far life. I grew up in Piplag, a small village in Gujarat, India. My first love was dancing, but at the end I fell in love with aerospace and defense technology to fulfill my parents' dream of seeing their daughters become either doctors, lawyers, or engineers.

Without my husband, I would not have been in a position to publish this book. We spent many sleepless nights debating each subject of this book. He led by example, striving to put out dents in the universe and never get complacent.

Finally, thanks to the person who helped me gain the freedom in my life, Mahatma Gandhi, who reminds me to "Be the change you want to see in the world."

Introduction

Practical Project Management for Engineers provides commonsense solutions to get projects rolling from contract award to delivery while using the Pareto principle to drastically increase productivity. The book walks the new technical project manager (PM) through step-by-step processes on how to deliver high-quality, robust products and services while strengthening one's team and customer relationships. While creating the book, Department of Defense (DoD), NASA, and Project Management Institute (PMI) project life cycle processes were compared and the best practices that have worked for our panel of experts on real-world projects are presented. This book is practical, not theoretical. We tell you "what you need to do on Monday." The project guidance from the DoD, NASA, and PMI and best practices from management and leadership gurus like Steven Covey, Peter Drucker, and Jim Collins have all been distilled into an easy-to-follow dialog. This knowledge will equip the new technical project manager to deliver quality products on time and on budget. Each chapter provides you with the most important activities and techniques that will aid any new project manager.

The last chapter of this book, *Tales from the Trenches,* was written by our panel of industry experts and gives insight to their lessons learned on real-world projects. In the real world, projects rarely meet schedules or avoid changes in scope.

By providing a practical outline of what is critically necessary for project success, efficient solutions are presented that will show you how to manage recovery plans and customer relationships to continue the future growth of your company and your career.

A project scope is defined by a set of requirements that define the characteristics of the desired products and services along with the conditions in which the project team will execute its work. It consists of a life span with a starting point and a defined end to deliver a quality product and/or service. The objective of project management is to deliver the products and services defined by the project scope within the constraints defined by the project's customer. The project's customer is the party that will benefit from the project's outcome. Project customers and sponsors can be internal or external to the PM's organization.

Effective project management requires specific knowledge, skills, tools, and abilities. Project management involves the application of these knowledge, skills, tools, and abilities to meet the project's requirements. Project management also requires a level of understanding of the environmental, operational, and technical aspects of the domain in which the project operates. For example, a person with expertise in project management processes and techniques supporting the construction of a high-rise tower requires an understanding of how buildings are constructed. Therefore, it is common for project managers to have some level of direct experience in the types of projects they support. This idea also holds true for complex scientific, engineering, or technical projects as the PM must be able to understand and communicate internally both with the scientific and engineering practitioners and externally to the slightly less-than-technical stakeholders.

The question arises for the technical professional transitioning to a project management or leadership position: What do I need to know right away to be successful in this new role? What do I tell my team "what you need to do on Monday"? The goal of this book is to help the new technical project manager to understand the most important aspects of project management to assist them in starting this new phase of their career.

Project success. The success of a project is measured by the convergence of the customer's expectations for the performance of the solution along with the allowable cost and schedule to deliver the solution (Figure I.1).

Project management processes. Projects are chartered to achieve one or more goals. Project management uses processes, tools, and techniques to achieve objectives in pursuit of these goals. This publication introduces the new technical project manager to the core processes we have identified for effective project management, including

- Communication management: Overcommunicate, listen first.
- Scope management: What does the customer expect?
- Schedule management: Who does what and by when?
- Requirements management: The product or service to deliver.
- Risk management: What could go wrong and what is the impact?
- Vendor management: Visit the vendor in person (rule number 1 for NASA project managers).
- Resource management: What you need to do on Monday.
- Cost management: Get paid.
- Configuration management: Everyone working from the same sheet of music.

Figure I.1 Cost/schedule/performance intersection.

- Quality management: Build quality and robustness into the product the first time.

Each process depicted in Figure I.2 is represented in a chapter within this book. While the processes within these chapters are presented in a sequential fashion, the processes are executed simultaneously. The results of one process will affect one or more of the other processes and the subject matter experts (SMEs) recruited to lead each of these processes are supporting members for the other processes.

You will need to understand

- What and how you must deliver a product and/or service before beginning any work.
- What you need to deliver a product and/or service on time and on budget per contract.
- Who in your project team will design, build, assemble, test, and deliver a product.
- How you are going to get paid: upfront full amount, incremental amount, or after delivery full amount.
- How you are going to manage recovery plans, since it is not an ideal world and things may fall behind schedule and under budget.

Communications management. Effective project managers are also effective communicators. Each project team member possesses a unique set of cultural and behavioral characteristics that drive how they communicate. Project managers must understand these individual behaviors to communicate effectively. Behavior profiling tests, such as the DISC test, aid in understanding these communication requirements so that the communications can be tailored to each audience [1].

Scope management. This is the process of writing a project *plan* that identifies clear roles and responsibilities, creating *processes* to remove nonessential work and show a clear path of how work is going to be done, and managing project requirements so quality *products* get built for the customer/end user. Bottom line: avoid scope creep by ensuring time, cost, and resources are validated before accepting a verbal or formal agreement.

Schedule management. This is the process of identifying the specific tasks to be performed and then developing a schedule for when the tasks are to be performed. Schedule management starts by developing a *work breakdown structure* (WBS) that defines the deliverables (hardware, software, services, and their subcomponents) that make up the project and creating work packages that break down the project into discrete activities to produce project deliverables. Care should be taken to validate that the total project SOW is included in the WBS prior to establishing the baseline for a schedule. The resulting project schedule is the road map used by the entire team to navigate through the project and to monitor progress.

Requirements management. Requirements are the single thread that goes through a project from conception through design, assembly, test, build, test and sell-off (customer acceptance). Requirements changes are inevitable. Project success relies on established, robust requirements management. Requirements management begins with analyzing customer "shall" statements and decomposing the elicited information into clear, achievable, verifiable, and actionable requirements.

Project requirements start with what the user really needs (not what the provider perceives that the user needs) and end when those needs are satisfied - Visualizing Project Management Forsberg, Moog, Cotterman [2].

Risk management. Risk is any uncertain event that could happen and the consequence of the event occurring. For example, foam coming off the space shuttle *Challenger's* external tank impacting the aircraft's leading edge (consequence: six astronauts losing their life), failure of a procured cast part impacting launch success (consequence: complete mission loss), and vague technical requirements left undefined prior to preliminary design review (PDR) causing rework (consequence: schedule and budget overrun).

Cheap or expensive missions can both succeed, but a rushed mission almost always fails. Failure is the norm: something goes wrong on almost every project. Space Systems Failure, David M. Harland and Ralph D. Lorenz [3].

Risk management is the formal process of identifying, analyzing, and monitoring risks:

- Risk is an uncertain event that could happen;
- Issue is realized risks that are now problems;
- Opportunities are future uncertainties that are favorable.

Vendor management. Successful vendors and suppliers enable successful projects. Selecting successful vendors requires identification of clear requirements and relevant contract details to ensure delivery of quality products on time and on budget. Vendor management ensures that the contract between vendor and supplier is clearly defined and includes the contract type, clear requirements, service agreements, key performance indicators, terms, and conditions. Establishing proper and accurate communications with vendors enables vendor success, which can lead to project success. This chapter provides important techniques such as holding a vendor visit and face-to-face kickoff, identifying and participating in key design reviews, defining clear acceptance requirements for all vendor deliverables, and maintaining professional relationships with the vendor throughout project life cycle and beyond.

Resource management. Project success relies on building and maintaining the best possible project team. Creating the best possible project team requires selecting the right team members. Building the project team by improving their capabilities, interactions, and the overall team environment leads to enhanced team performance. Managing the project team involves tracking team member performance, providing feedback, resolving issues, and managing changes within the team. Project managers must understand budget and scope before selecting a team. By selecting, building, and maintaining the best possible team, PMs can recognize savings in cost and schedule while achieving improved product and service quality.

Cost management. All projects require the project manager to determine the cost to complete project work and manage the work within this initial planned cost. Earned value management (EVM) is a method used to track current project performance and forecast future performance. EVM integrates (planned versus actual) scope, schedule, cost, and risk management. The best way to manage project cost is utilizing an earned value management system (EVMS) as it provides insight to key questions of cost management:

- Schedule performance index (SPI): Are we ahead or behind schedule?
- Cost performance index (CPI): Are we under or over budget?
- Estimate to complete (ETC): What will be spent on remaining project?
- Estimate at complete (EAC): What will be spent on whole project?

Configuration management. Configuration management (CM) is the systematic control of changes to the identified configuration for the purpose of maintaining product integrity and traceability throughout the product life cycle. CM documents all the versions, builds, and baselines of all project configuration items (CIs) and maintains a record of all changes to CIs throughout the project life cycle. CM ensures all information (both formal and informal) is tracked to ensure accurate version control of the information and that any verbal agreements are properly documented.

Quality management. Quality management (QM) is the practice of designing and building products right the first time. QM is more of a process than a product. It's a continuous improvement process to enhance future products and services. Quality assurance prevents defects on the process used to make the product (preventive process) and quality control detects defects in the finished product (corrective process).

- Quality assurance: Doing the right things the right way.
- Quality control: Ensuring the results of what you've done are what you expect. (Is product conforming or nonconforming?)

Organizational process assets. The first piece of good news for the new PM is that mature organizations have a set of standard processes, tools, practices, and templates available to aid the project manager. These assets may include training, example products, knowledge bases, or process area mentors. The common term used within the project management community and this publication is organizational process assets (OPAs). The first tip for the new PMs is to find, understand, and use their organization's process assets. Each chapter in this book identifies the OPAs that should be available to the new PM.

Organizational systems. Mature organizations possess a set of systems and data also available (if not required) for use in executing project management activities. These systems include the organization's enterprise resource management and planning systems (ERP or ERM), which are the authoritative source for all information pertaining to human, capital, facility, and financial resources. Other examples of these systems are enterprise portals and project management tools. Tip number two for new PMs is to identify their organization's enterprise systems and the parties with authority to access and use them.

Project origination. Projects exist to deliver solutions to an organization's need to change one or more conditions within the organization. These needs are identified by the parties involved in the leadership and execution of organization's business or mission. Needs come in the form of a gap in capability, a desire to improve performance, a need to address risk, or to modernize existing capabilities. Organizations use formal and informal processes to define the needed change and the range of options for potential solutions. Documentation created by the analysis performed leading up to a new project can be in the form of

- Business or mission need statement;
- Initial capabilities description document;
- A business case document.

Projects are initiated by authorizing the use of resources to define, develop, and deliver the required change. In many cases, these steps are performed in advance of assigning a project manager to the yet to be authorized project.

The process of authorizing new projects should result in some form of charter document that defines the high-level scope, objectives, authorities, and resources for the project. These documents may come in the form of a project charter, task statement, and/or a formal contract.

Importance of planning. Projects are chartered to achieve one or more goals for the customer, and the project management uses processes, tools, and techniques to achieve objectives in pursuit of project goals. Planning is the thought-based process of identifying the activities and conditions required to achieve a goal or objective. The overarching plan for the project manager is the project management plan (PMP), which contains the integrated set of plans to achieve the objectives associated with each process [4]. Therefore, each process area depicted in Figure I.2 requires a level of planning to achieve its objective and an understanding of how each process area supports the other PM processes.

The sample project. A valuable aspect of this book is the use of real-world examples to demonstrate the activities and resulting outcomes associated with the processes described in each chapter. The project selected for these examples is a less-than-real-world project to create the Illudium Q36 Space Modulator depicted in the Warner Brothers Bugs Bunny Series [6]. This project is strictly a satirical method to demonstrate the typical data created for a technical project. However, the data and the graphic representations are drawn from real-world project files created by the authors to plan and monitor the Q36 project.

Figure I.2 Core project management processes.

References

[1]   Manager Tools, "Using DISC for Effective Communications," November 2018, [online], managementtools.com.

[2]   Fostburg, K., H. Mooz, and H. Cotterman, *Visualizing Project Management: Models Frameworks for Mastering Complex Systems,* Third Edition, Hoboken, NJ: John Wiley & Sons, 2005.

[3]   Harland, D. M., and R. D. Lorenz, *Space Systems Failure: Disasters and Rescues of Satellites, Rockets and Space Probes,* Chichester, UK: Springer, 2005.

[4]   Project Management Institute, Inc., *A Guide to the Project Management Body of Knowledge: PMBOK Guide,* Sixth Edition, Newtown Square, PA: Project Management Institute, Inc., 2017.

[5]   International Institute of Business Analysis, 2018 [online], www.iiba.org.

[6]   Jones C. M., (director), *Haredevil Hare* [film], United States: Warner Brothers Cartoons and Vitaphone, 1948.

CHAPTER 1

Communications Management

According to research performed by Iman Attarzadeh and Siew Hock Ow of the University of Malaya, over 50% of project failures are due to some form of communication breakdown [1]. This statistic indicates that there may be no greater single subject with more influence on project success than communications. However, this same study also found that while many executives understand the importance of effective communications, only a fraction of project managers recognized communications as a core competency [1] (see Figure 1.1). Given the role of communications in project success, one would expect to see a high level of formal training and materials focused on the topic of effective communications. However, one may be shocked to learn how little formal training or discussion occurs on communications at an enterprise level. This chapter introduces the new technical project manager to a set of tools and techniques found to be effective when communicating within or on behalf of the project team.

Jerry Madden, associate director of the Flight Projects Directorate at NASA's Goddard Space Flight Center, collected a series of observations and best practices for successful project management. Mr. Madden collected these gems of wisdom over several years from various unattributed sources. NASA's *ASK Magazine* published these practices as "100 Lessons Learned for Project Managers" in 2003 and these "100 Rules for NASA Project Managers" as unofficial guidance to all project teams [2]. The following are a few of these rules that focus on communications:

NASA Guidance for Project Managers, Communications

Rule 16: Cooperative efforts require good communications and early warning systems. A project manager should try to keep his partners aware of what is going on and should be the one who tells them first of any rumor or actual changes in plan. The partners should be consulted before things are put in final form, even if they only have a small piece of the action. A project manager who blind- sides his partners will be treated in kind and will be considered a person of no integrity.

Rule 17: Talk is not cheap; but the best way to understand a personnel or technical problem is to talk to the right people. Lack of talk at the right levels is deadly.

Rule 18: Most international meetings are held in English. This is a foreign language to most participants such as Americans, Germans, Italians, etc. It is important to have adequate discussions so that there are no misinterpretations of what is said.

Rule 19: You cannot be ignorant of the language of the area you manage or with that of areas with which you interface. Education is a must for the modern

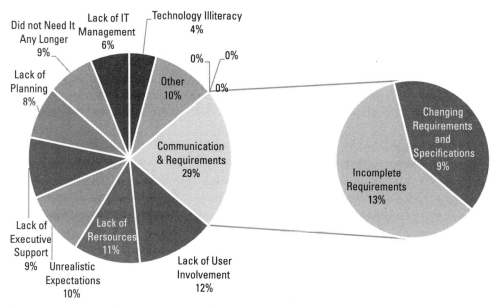

Figure 1.1   Project criteria for success or failure. (From [1].) (Reprinted with permission of Manager tools, LLC © [5].)

manager. There are simple courses available to learn computerese, communications and all the rest of the modern "eses" of the world. You can't manage if you don't understand what is being written (*Source:* [3]).

Know the Audience

Merriam-Webster's dictionary defines communications as "a process by which information is exchanged between individuals through a common system of symbols, signs, or behavior" [6]. When communicating, it is imperative to understand the technical, social, and cultural perspective of the person one is communicating with.

Project managers frequently communicate with a broad range of stakeholders, each with different perspectives and technical backgrounds. Complex technical information shared with an engineer may frustrate or threaten a person with a less technical background. At the same time, oversimplifying an issue when communicating to a highly experienced audience may erode confidence in the speaker.

Additionally, each person with whom a project manager (PM) communicates has different personality traits and behavioral characteristics that affect each person's communications style. Where some people can interpret complex topics with minimal information, others require introductory information before building up to a complex topic. Where one person comprehends better with graphical representations others, prefer to see the raw data.

Behavior and personality assessments provide project leaders with the ability to understand the unique characteristics of each person on the project team. Available under several variations, these assessments create a behavioral or personality profile for each person that can help a communicator to understand his or her audience. The DISC assessment methodology is based on the research into human behavior performed Dr. William Moulton Marston. Marston's theory stated that people behave based on their emotions and their emotions are based on their perception of the environment within a given situation. These perceptions can be positive or negative:

- *Dominant:* Perceives oneself as more powerful than the environment and perceives the environment as unfavorable;
- *Influential:* Perceives oneself as more powerful than the environment and perceives the environment as favorable;
- *Steady:* Perceives oneself as less powerful than the environment and perceives the environment as favorable;
- *Conscientious:* Perceives oneself as less powerful than the environment and perceives the environment as unfavorable;

These assessments typically ask the person to respond to a series of questions and scenarios. The responses are them compiled to create a profile for the person. The profiles catagorize each person as being either *people focused* or *task focused*, and if the person is *introverted* versus *extroverted*.

Effective project managers (PMs) tailor communications to each audience by using language, content, and diction suited for the individual. Tailored communications are genuine, not fake. Tailored communications avoid manipulation of the audience by limiting the use of complex terms, references or acronyms while including the level of detail required to communicate a topic to the individual audience. By understanding a person's natural communication style and adjusting to it, one can more effectively communicate ideas [4].

Figures 1.2 and 1.3 provide quick reference guidance for using DISC information to help shape communications to the DISC profile of the audience.

It is a wise investment to have everyone on the team take a DISC profile assessment and discuss the meaning of the results. Even better, providing a day-long communications course based on DISC can enable the entire team to improve its communications [7].

Effective Communication Media

Not all communications media are equal in effectiveness. Each type of communications has its advantages and limitations in supporting the exchange of ideas and information. Factors such as body language, eye contact, facial expressions along with the tone and inflection of voices all influence the level of attention, collaboration, and information transfer in a conversation.

Collaborative face-to-face communications in which the participants can exchange ideas using a white board is the most effective communications medium followed by that of basic face-to-face communications. Table 1.1 provides the most common communications media each rated for effectiveness. The effectiveness scores provided reflect each medium's expected effectiveness as compared to that of face-to-face. As the table demonstrates, each step down from face-to-face communications results in the risk that communications will be less effective.

How To Use The DISC To Be More Effective Every Day			
High S's – Steadiness			
How You Can Spot Them:		**What They Want From Others:**	
How They Talk:	**What They Do:**	High S's like others to be relaxed, agreeable, and cooperative, and to show appreciation	
• Make small talk	• Photos of relationships out		
• Ask how questions	• Consult others	**You Should Try To:**	**Be Ready For:**
• Ask vs. Tell	• Friendly functional work area	• Be logical and systematic	• Friendly approach to others
• Listen more than talk	• Casual relaxed walk	• Provide a secure environment	• Resistance to change
• Slow, steady delivery	• Patient, tolerant	• Tell them about change early	• Difficulty prioritizing
• Reserved w/ opinions	• Service oriented	• Use sincere appreciation	• Difficulty with deadlines
• Lower volume	• Embarrassed by recognition	• Show how they're important	
• Warmth in voice	• Subdued clothing	• Let them go slow into change	
• Use first names			
How To Manage Your High S's			
You Can Help Them Learn:		**They May Want From You/ Your Organization:**	
• Openness to change	• Short cut methods	• Status quo	• Security
• Self-affirmation	• Effective presentation skills	• Private appreciation	• Time to adjust to changes
• How to make their accomplishments known	• Believing their successes are worthwhile	• Happy, calm relationships	• Listening
		• Standard procedures	• Sincerity

High C's – Conscientious			
How You Can Spot Them:		**What They Want From Others:**	
How They Talk:	**What They Do:**	High C's like others to minimize socializing, and give details; they value accuracy and attention to detail	
• Ask Why questions	• Focus on task and process		
• Ask vs. tell	• Orderly	**You Should Try To:**	**Be Ready For:**
• Listen more than talk	• Meticulous	• Give clear expectations/deadlines	• Discomfort with ambiguity
• Not a lot of reaction	• Precise, accurate	• Show dependability	• Resistance to vague information
• Slower speech	• "Sterile" work area	• Show loyalty	• Desire to double check
• Lower volume	• Time conscious	• Be tactful and reserved	• Little need to be with others people
• Prefer to talk vs. writing	• Hard to read	• Honor precedents	
• Get to point but like to talk	• Diplomatic	• Be precise and focused	
• Precise, detailed speech	• Want to be right	• Value high standards	
How To Manage Your High C's			
You Can Help Them Learn:		**They May Want From You/ Your Organization:**	
• Tolerance of conflict	• Acceptance of others' ideas	• Clear expectations	• No sudden changes
• To ask for support	• Tolerance of ambiguity	• Limited exposure	• Personal autonomy
• Group participation skills	• Acceptance of their limits	• Business-like environment	• Chance to show expertise
		• References & verification	• Attention to their objectives

Figure 1.2 Uisng DISC to shape communications [4]. (Reprinted with permission of Manager tools, LLC © [5].)

In the day-to-day rush to complete tasks, it is tempting to use the communications medium that is faster and easier for the sender. However, methods like email, phone and chat are less effective. Email and other asynchronous communications can be used in cases where the risk of miscommunication is low. However, for complicated or potentially confrontational discussions it is best to use a method that increases the data transfer relative to tone, body language and interpretation. Whenever communication starts to break down, moving up the hierarchy in Table 1.1 is the best way improve the situation.

On exact language: English is a great language with all the words needed to convey complex ideas. One does not need to invent new words or use the latest trendy words that don't have self-evident meaning to communicate effectively. Use simple words when simple words will do. Explain acronyms the first time you use them and don't always use them assuming everyone knows what they mean.

For example: use the word *perception* not *optics* when describing one's interpretation, discernment or sensations. *Optics* are pieces of glass ground into lenses.

| How To Use The DISC To Be More Effective Every Day |

High D's - Dominance			
How You Can Spot Them:		**What They Want From Others:**	
How They Talk:	**What They Do:**	High D's like others to be direct, straightforward, and open to their need for results	
• Ask What Questions	• Task Focus, Results Oriented		
• Tells vs. Asks	• Impatient	**You Should Try To:**	**Be Ready For:**
• Talks More Than Listens	• Direct, Forceful	• Communicate briefly/to the point	• Blunt/demanding approach
• Go Right to The Issue	• Willing to Get in Trouble	• Respect their need for autonomy	• Lack of empathy
• May Be Pushy, Even Rude	• Time Conscious	• Be clear about rules/expectations	• Lack of sensitivity
• Fast Speech	• Good Eye Contact	• Let them take the lead	• Little social interaction
• Authoritative Tone of Control	• History of Achievement	• Show your competence	
• Use Acronyms, Short Sentences	• Can Rely on Gut Feelings	• Stick to the topic	
• Open w/ Opinions	• Maverick	• Show independence	
How To Manage Your High D's			
You Can Help Them Learn:		**They May Want From You/ Your Organization:**	
• Identifying with others	• Ways to pace themselves	• Power and authority	• Results
• Empathy for others	• Relaxing	• A promotion	• To know the bottom line
• More logic, less gut	• To be approachable	• Prestige	• Freedom from details
• Listening skills	• Complimenting others	• Big challenges	• Direct answers
• To "soften" body language	• To ask more questions	• Authority to make changes	• Flexibility

High I's - Influence			
How You Can Spot Them:		**What They Want From Others:**	
How They Talk:	**What They Do:**	High I's like others to be friendly, emotionally honest, and recognize the I's contributions	
• Ask who questions	• Animated		
• Tell vs. ask	• Lots of facial expression	**You Should Try To:**	**Be Ready For:**
• Make small talk	• Spontaneous	• Approach them informally	• Attempts to persuade/influence
• Go off on tangents	• Laugh out loud	• Be relaxed and sociable	• Need for the spotlight
• Use stories or anecdotes	• Stylish dress	• Let them tell you how they feel	• Over-estimates self/others
• Faster speech	• Shorter attention span	• Keep the conversation light	• Over-selling ideas
• Express their feelings	• Warm	• Provide written details	• Vulnerable to feeling rejected
• Share personal emotions	• May approach you closely	• Give public recognition	
• Exaggerate		• Use humor	
How To Manage Your High I's			
You Can Help Them Learn:		**They May Want From You/ Your Organization:**	
• More control of time	• Organization	• Popularity	• Casual warm relationships
• Objectivity	• Sense of urgency	• Visible rewards	• Freedom from details
• Emphasis on clear results	• Analysis of data	• Public recognition	• Approval And friendliness

Figure 1.3 Using DISC to shape communications [4]. (Reprinted with permission of Manager tools, LLC © [5].)

Engineers respect precision and precision in communications aids in good thinking when making complex decisions. When possible, use exact figures. For example, using exact words adds precision to the conversation while demonstrating competency in one's profession.

Email Use

Since its wide spread introduction in the early 1990s, email has become the default tool utilized for both informal and formal communications. The ease and immediacy of email is alluring, but these traits are often what prevent it from being a truly effective communication medium.

Email is one of the few time-wasting activities that are sanctioned within today's corporate culture. Email can be a tool to accomplish certain communication priorities. Few employees are assigned an annual performance goal for answering

Table 1.1 Communication Medium Effectiveness [4]

Communications Medium	Attributes	Effectiveness
Face-to-Face with whiteboard	Audio, tone, inflection, facial expression, body language, eye contact, full attention	1.0
Face-to-Face	Audio, tone, inflection, facial expression, body language, eye contact, full attention	1.0
Video Conference	Reduced: body language, facial expression, eye contact, attention	0.7
Phone Conference	Loss of: facial expression and eye contact, reduction of tone and inflection, reduced attention	0.5
Email Exchange	Loss of: all visual cues, audio, tone, inflection, expression. Increased potential for misinterpretations and language barriers	0.3
Chat/Text	Same as email with even less data exchanged. Less formality.	0.2
Written Documents	Loss of: all visual cues, audio, tone, inflection, expression. Increased potential for misinterpretations and language barriers. Risk for lost attention.	0.3

http://www.ambysoft.com/surveys/practicesPrinciples2008.html

email. Email can be a tool to use when performing project duties but responding to email all day is not an acceptable excuse for failing to achieve project objectives and responsibilities. When reading and responding to email becomes the principle focus of one's work day, it may be time to re-evaluate the use of email as a communications tool. The following are some thoughts and recommendations to help your emails stand out as beacons of beauty, clarity, and effectiveness.

Avoid distractions: It is a common misconception that email must be read and responded to immediately. Reading emails while working on other tasks is the equivalent watching television while holding a conversation with a friend or loved one. Neither the TV show nor the other person's communications are fully interpreted, and important details are lost. Effective communicators schedule time frames throughout the day to focus solely on email. They also limit these intervals to no more 30 or 45 minutes each and limit these sessions to no more than 3 times a day which equates to 1½ or 2 hours out of an 8-hour day. For example; email intervals at 8am, 12:30pm, and 4:30pm allow one to clear their inbox in the AM and after the lunch break before starting productive activities. The last session can aid in identifying scheduling priorities for the next day. A good way to maintain a disciplined approach to email time consumption, is to turn off all desktop and audio alerts for incoming messages.

Email Distribution

The following provides some guidance for the appropriate use of various email distribution options.

To: Obviously the person who is the intended recipient. Usually, this should only include one or a few people. If anyone is being tasked, or asked a question where their response is expected, they MUST be included in the "To:" field.

Table 1.2 Exact Words

Exact Words	Ambiguous Words
Profits are up 5%	Profits barely moved
I'm out Monday through Friday Christmas week	I'm out this holiday
At 70% applied load, we experienced a major structural failure in the strut fitting	The hardware blew up

Table 1.3 Effective Subject Lines

Poor Subject Line	Effective Subject Line
Starbucks?	Do you want to meet at Starbucks at 2pm?
Re: Do you want to meet at Starbucks at 2pm?	Subject: Re: Do you want to meet at Starbucks at 2pm? [yes, see you there]
"yes, see you there" in the email body	
PTO	PTO for John Hamilton this Friday from 12-4pm

CC: (from history is short for carbon copy. Often incorrectly believed today to mean courtesy copy) This is usually used as a CYA (cover-your-ass) tactic. The recipient of a cc correspondence is NOT obligated to read the email. It is often used to allow people to be as informed as they want to be and have a record of events and attachments if they need it in the future. The higher rank a person is in an organization, the more likely they never read cc'd email unless asked to dig them up. You should only cc people when it's obvious they should be informed, or they have asked to be included in the email discussion. cc'ing your whole organization (or manager) for every email is both ineffective and will make everyone hate you for being foolish and egotistical enough to think everyone should care about all your email. Cc'ing a manager to 'escalate' pressure on the "To:" recipient should almost never be done. If you need something that badly, pick up the phone or drop by their desk to ask in person. A little eye contact will get you more compliance than the passive-aggressive threats of having your manager beat up his manager. We're adults, act like one.

BCC: (blind carbon copy) Never use BCC to a person. I use BCC for archiving purposes only. The people you send email to should be given the courtesy of knowing who you sent the original email to. There is an assumption of integrity in the distribution lists. You only need to have someone whom you bcc'd "reply-to-all" once to experience the fallout of your clever little IT trick unraveling. Just don't do it.

Avoid reply-all, almost never request a reply-all thread. I've seen this done once, and it was stupid. It's what happens when people don't know any better.

Effective Subject Lines: In some cases, an effective subject line can be the entire email. By placing specific information in the subject line allows the recipient to be efficient in their email management.

These examples demonstrate how a little more information in the subject line can result in a faster response from the recipient. This technique should not be a replacement for professionalism within the body of the email. In the case of the request for time off, one is still be expected to include the required formality in the body of the message, but the subject line can fully describe the essence of the conversation.

BLUF–Bottom Line Up Front: Emails should not require the recipient to hunt for the point in the communication. Like using the subject line to provide insight to the content of the email, opening an email by first stating the point, question or message (the bottom line up front) can also improve the speed of email communications. When asking a question, it can help to do it first and then explain the backup information. The recipient may be able to simply answer the question without reading all the backup information. Email is not storytelling, it is business communication. Effective PMs put long emails aside in favor of shorter, more concise emails and storytelling is a good way to have your emails ignored.

Use Short Signature Blocks: Every email should have a signature block. Signature blocks should contain the contact information for both electronic and voice communications so that the recipient can respond in the manner of their choice. Signature blocks should include only formal and professional content. Jokes, quotations and other nonessential information should be excluded from signature blocks. All signature blocks should include the following:

Name
Title
Company
Office Phone
Cell Phone

Avoid using the Confirmation and Priority Features: Using these features in modern email systems does little more than create a high level of discontent with recipients. These features make people feel attacked and wonder why the sender thinks his email is so important that has have to know when it is delivered. A phone call is more professional way to verify that a critical email has been received and read.

The priority feature may be the most overused and ignored email feature of all. Some email users have a habit of tagging every email as *High Priority*. The result of this overuse of the priority flag is that people have become desensitized to seeing red flag on the email and ignore it. Rather than using a priority flag to convey importance one may find it more effective to have a person-to-person conversation.

Use email efficiently: Limiting the time spent on emails may require prioritizing which messages get attention. The priority may be those emails where the sender practices the effective email habits provided within this chapter. Messages in which the reader is directly addressed (in the To line) should take priority over those in which the reader was carbon copied (Cc) line. Short and to-the-point emails require far less time to process than those with long, drawn out narratives. A good technique is to move emails that require more time into a folder for when time allows later in the email session. Emails that require no action but contain information that may be of use later should be tagged and filed to aid in retrieval. Lastly, emails that require no action and are of no reference quality should be deleted.

Often with email and other collaborative work, one can get stuck by overthinking their responses. The desire to respond thoughtfully can add an unacceptable delay in responding. Effective communicators seek to provide an acceptable response instead of a perfect response. For example: An engineer receives an email requesting an estimated cost for an item. The engineer knows the cost to be between $100 and $120. He also knows it will take four hours to refine this to a more exact amount. The engineer responds by stating: "I am pretty sure the answer is between $100 and $120, but with four hours of work I can refine it to a more exact value. Please advise if you require an exact value." By giving this initial estimate right away, the engineer can move onto the next email in his list and can wait to see if a more exact value is required.

Avoid Distribution Wars: Almost all have experienced email wars where each new response to an email comes with more people added to the distribution. Professionals see this flawed tactic a mile away and do not respect those who use the tactic. Never use email for negative conversations, professionals speak to the person directly and in private.

Let It Burn Policy: The people who win email wars are the ones who don't get involved. Effective professionals understand the importance of de-escalation when they are included in an email war. These professionals either ignore the conversation and let it die out or when forced to engage, they limit the distribution to only those critical to the conversation.

Another good leadership practice is to call the individuals directly to help defuse the situation. A friendly, conversational tone and willingness to help usually calms things down.

Abbreviations: Some managers have adopted the use of a short list of acronyms when sending brief *internal* messages. Acronyms such as EOM (end of message), FYI (for your information) and NRN (no response needed) can help to add details to the subject line. An example of which might be:

Subject: Snow has caused the factory to close, no work today [NRN]

Phones and Voice Mail

Table 1.1 shows that voice communication is more effective than email and therefore it should be used whenever possible. When face-to-face engagements are not feasible, the phone can be the next best method. Voice mail can also be an effective communications method when the called party is unavailable.

Placing Calls: When placing calls, start the conversation by identifying themselves to the answering party. After exchanging pleasantries, the professional will explain the reason for the call, how much time the conversation may take and if it is a good time to have the conversation. For example: "John, this is Eric at Raytheon, do you have 10 minutes to discuss the budget in the proposal, or should I call you back another time?"

Accepting Calls: The importance of curtesy and professionalism also applies when answering calls. The absence of a face to face presences is not an excuse to multi-task or allow distractions. If the timing does not support a lengthy conversation at that time, it is acceptable to offer to call the person back when the proper time and attention can be afforded.

Respond to voicemails: Like email, voicemail responses should be within 24 hours unless it is a weekend or holiday. The one exception to this weekend and holidays rule is when travel or other conflicts will preclude calling during business hours. Reaching the other person's voicemail can still be an opportunity to move the conversation forward. The caller can leave a detailed message, respond to the initial voicemail question or provide a time when he or she will be available to receive a returned call. For example; Eric has called John about meeting for lunch, he gets John's voicemail and provides the following "John, this is Eric. Let's meet for lunch on Tuesday at Max's at 11:45". When John returns Eric's call he too gets Eric's voicemail and leaves the following: "Eric, this is John. Lunch sounds good, I'll see you at Max's at 11:45 on Tuesday". In this example the two phone calls were missed by their recipients, but the key information was successfully exchanged between the two parties.

Effective Meetings

Meeting Preparation

Discussions in many meetings seem to wander aimlessly for most of the meeting with all decisions made in the last 15 minutes. Effective leaders come to meetings prepared with the decisions targeted for each meeting. The meeting discussions are then focused on these decisions throughout the meeting, effectively eliminating the need for the last 15 minutes. Being prepared not only improves the effectiveness of meetings, it can also shorten meeting times.

No Meeting Wednesday

A common complaint among project teams is that they spend too much time in meetings. The fact of the matter is that very little work is accomplished while attending a meeting. Effective managers understand the need to strike a balance between the need to collaborate and the need to focus attention on the project's work. A common technique in organizations is to have a day each week or certain hours in each day where meetings are prohibited. This policy demonstrates respect for the team's time and enables people to have some control over their schedules.

> *A multi-billion-dollar project I was supporting for a major aerospace prime was running years behind and having technical challenges. Out of frustration of having so many meetings yet making so little progress in the way of design closure and drawing release, the Program Manager instituted a "No Meeting Wednesday" policy. No team member was authorized to attend any meetings on Wednesday. Wednesday was the day dedicated to making work progress. I don't know why an organization would have to be in crisis to implement this. I think it keeps the focus on actual task work each week.*
> —Eric J. Roulo, Aerospace Structures Consultant

Use WebEx

WebEx is a great tool to help people collaborate across locations. You should make a habit of using the entire capability of WebEx for conference calls. This would include, recording the meetings, using video conferencing, screen sharing, and use of the participant window.

Log into the web interface. It allows you to share screen content, video, and see who's talking and who's online. There is no need to have a roll-call on a modern conference call, you should be able to see who's logged in based on the participants window. The performance of WebEx is excellent when sharing engineering or visualization programs that rotate 3D models or render.

Direct, Informal Communication

The single most important behavioral trait to aid in effective communication is to foster direct, informal communication between all members of the project. The need for effective communications includes interactions with the project's suppliers, vendors, and customers. Establishing a weave of intersecting, informal lines of communications helps prevent critical data from being siloed or silenced due to an overly strict command and control information distribution approach. This openness may ruffle some managers who wish to control the project with a tight grip on personal interactions and information sharing, but preventing information sharing through informal communication often leads to long-term problems. Effective teams do not ask for permission to execute, and effective team members should not require permission to communicate and collaborate. Professionalism must of course be maintained, and team members cannot agree to verbal orders from customers or authorize changes to a vendor's scope of work.

Allow Back Channels

Creating back channels can help when things get rough. It's important for projects and the organization in general to have good working relationships with its customers at all levels. Many defense contractors try to limit the communications between sub-contractors, vendors, and customers. There are many legal reasons why this level of formality must be enforced. Some information associated with the terms and conditions of a contract between buyer and seller must always be protected.

There will be times when information needs to travel quickly and informally to assess project options. For example: if the requirements are going to change, it could be beneficial for the engineer to call up his counterpart at the customer and ask, "how much of an impact would it be if we asked for some requirements relief on the voltage regulator?" The customer may share that they don't care at all, or they have already built downstream hardware that requires the current specification, or they might be able to accept a change temporarily, but after that, they will start building hardware that requires the original parameters. In this example, if the engineer had waited to go through formal channels, the request might have occurred when it was too late for changes. This conversation does not remove the need for formal change control. However, asking first helped the engineer to evaluate the feasibility of a change before starting the formal process. It is good practice to set down clear ground rules for the type of collaboration between the project team members that is and is not appropriate.

Heads Up

Eye contact is important for good communications. In Western cultures, looking one in the eye when conversing is a sign of respect and implies to the speaker that attention is being applied to the conversation.

Be Accessible

Good leaders are never too busy for their teammates. When in the office, it is important to regularly look up from one's desk or the phone in order to signify availability. Working behind closed doors or while wearing headphones also signifies unavailability. Publishing open office hours is a good tactic for striking a balance between personal productivity and team accessibility.

One approach to increasing one's accessibility is to take lunch breaks in a public location such as the break room or cafeteria. It may help the PM to be more connected with the team by having lunch with people and avoiding eating alone. Lunch is one of the easiest ways to implement informal communications channels, find new mentors and coaches, make friends, and hear about the goings-on around the company.

The Importance of Templates

Consistency in the format, content, and structure of routine communications aids in the efficiency and effectiveness of communications. Templates are an effective method to standardize communications while providing clear guidance to the project team. Standard project templates should be created for the common forms of communications required by the team. The following includes some key examples of communications templates all of which can be accessed and downloaded at rouloconsulting.com/templates (see Figure 1.4).

Trip Reports

The activities that transpire each time a member of the team ventures out to meet with external parties should be documented and shared with the project team. This practice serves to keep all parties informed while improving the timelessness of detecting potential issues.

Weekly Status: Four Corner Charts (Quad Chart)

Periodic information sharing sessions are a good way to help increase awareness across the team. One such information sharing technique is to have each team member prepare and present a single page status report during the weekly team meeting. This approach provides a brief 5-minute platform for everyone to explain what they are working on, what their challenges are, and to seek help from the team. These information sharing sessions are in many ways a teaching moment. During these sessions, other team members can share their expertise in the problem area, the presenter gets experience presenting in public, and all attendees learn about the

TRIP REPORT
DOCUMENT NO. SMC-TR-2011-0001

WHO: [CLICK HERE AND TYPE NAME] *WHO WAS MET*

WHAT: [CLICK HERE AND TYPE NAME] *WHAT WAS DISCUSSED*

WHEN: [CLICK HERE AND TYPE NAME] *DATE OF THE MEETING*

WHERE: *WHERE DID THE MEETING OCCUR*

WHY: [CLICK HERE AND TYPE NAME] *WHAT INSTIGATED THE MEETING*

WHAT'S NEXT: *WHAT ACTIONS NEED TO BE TAKEN NEXT*

Business trips consume large amounts of both capital and resource time. For this reason it is desired to prepare in a standard and effective way for the trip to maximum potential benefit and also to report outcomes from the trip to best capitalize on the investment. The trip report is the mechanism to capture information collected during the business trip, record lessons learned, and create action items to follow up to convert the business trip in continued business or future contracts.

Executive Summary:
In one or two paragraphs briefly describe the trip outcomes and future actions to take to move the relationship toward a mutually beneficial and lucrative business relationship.

Description of the Trip:
Transcribe the trip notes and critical details here. Be sure to include all personal introductions and, business cards exchanged, future linked in connections, etc. How many ways can be build business relationships from this trip?

Conversation Details:
Describe any conversations that were of interest where critical details were revealed. Be especially mindful of challenged that were described by the client where SMC could provide solutions or value. Look for discussions of future business outlooks and manpower, software, and tools and solutions that SMC might be able to provide.

Follow Up Actions:
List the actions, who will complete them, and the estimated completion date for follow up to move the meeting toward getting a proposal or contract signed. All actions should be input into Hirerise or Basecamp before uploading the final document into basecamp and the server.

Action	Assignee	Estimated Complete Date

Lessons Learned:
Describe what worked, what didn't work, what could be improved on. List restaurants reviewed or visited, costs, ambiance, etc.

Of Historic Note:
List interesting details that may be of historic interest for future meetings or general background.

Create TDL for trip report completion
1) Enter data into all the sections that are applicable
2) Remove the instruction text in italics

The Structural Mechanics Corporation
www.structuralmechanics.com   Page 1 of 1

Figure 1.4   Trip Report template [6].

various aspects of the project. These presentations also help the team to prepare for upcoming customer reviews.

80-20 Report

Eighty-twenty reports are a modified version of a root cause analysis method originally created by Toyota based on the pareto principle. An example of pareto principle applied to project management is that most of a task value (~50-80%) is achieved after only a portion (~20%) of the work is performed.

Using this theory, one could then assume that a good time to proactively review a task's risk is after 20% of the work is accomplished

The purpose of this review is multifaceted:

- Force the analyst to see a potential problem while time allows for correction,
- Defend against spending 80% more time on an analysis project or design task that will likely result in an unacceptable design configuration,
- Provide management a reasonable risk assessment early in the analysis or design effort,
- Provide a gate, which allows a midcourse correction before the project veers too far off track,
- Provide a teaching tool for less experienced people
- Pprovide lessons learned and templates for future work.

The report is meant to be relatively informal but provide enough detail to clearly describe the analysts thinking process for use later by others. The minimum discussion topics would be:

1. The problem description (copied from other required SOW or final reports)
2. The desired analysis result (what is the task objective?)
3. The most efficient analytical approach or approaches used to bound the problem
4. The results of that analysis
5. Recommendations

Template Evolution

Templates are not meant to constrain project team members. They are intended to help execute their communications efficiently and effectively. Templates provide people with a minimum set of information expectations in the desired format and are meant to evolve as time and methods change. Parts of the templates that proven to be ineffective should be rewritten or discarded. Revisions and better examples provided from within the team should be incorporated. For instance, the current template for the four-corner report in Figure 1.5 [6] incorporates all the good ideas and clever ways to report status that have been collected over years of experience.

Four Corners Chart

Eric J. Roulo - December 14, 2012

Accomplished this Week	Planned for Next Week

• NDE Report Support (meetings/planning)
  • (losing NDE buy-in)
• Octaweb additional load cases processed
  • No negative margins for TPA/engine out
  • Scripts for TFD and crack growth 50% complete
  • Xfer to Tjepke slipped
• Computation Engineering 2 Python (WG#3) completed
  • First script release due 12/19!
• Released Polina's Neuber memo
 	
• Consulting support for:
  • Erik Dambach (super draco testing)
  • Florian Kapsenberg (NX glue)
  • Polina (Crew methods approach)
	• Fairing knuckle and nose margin support (first priority)
	
	• OctaWeb Analysis Support (A)
	  • Allowable crack length plot (12/11 -> 12/21)
	  • Xfer margin calculator to Tjepke (12/11 -> 12/21)
	  • Process additional load cases run
	• Incorporate edits for Sandwich Core Margins memo (B)
	• Fastener Compliance Worksheet (from last month) (12/21)
	
	• NDE oversight support (weekly meeting)
	• Python WG#4
	• Review Polina's memos on Coffin-Manson
	• Continue general support of structures team

Schedule and Budget	Issues / Help Needed
34% of PO 181988 spent (540 hours remain)	Fairing review will continue over weekend and put other priorities at risk for next week completion I will work remotely 2-3 days next week to make more progress on lingering tasks Updated Holiday December 21st 2012 - January 7th 2013

Figure 1.5   Example Four Corners chart [7].

Conclusion

Good communication habits are critical for project success. Communication breakdowns are the number one category of project killers and deserve significant attention by management and team members. Using the methods described in this chapter can help the project stay on track and conserve resources. The importance of having a lot of informal communication channels and overcommunicating on every level of the project team can't be overemphasized enough.

References

[1] I. Attarzadeh and S. Oh, "Project Managment Proactices: The criteria for success or failure.," *Communications of the IBIMA*, vol. 1, p. 8, 2008.

[2] J. Madden, "100 Lessons Learned for Project Managers," *ASK Magazine*, October 2003.

[3] Jerry Maden and Rod Stewart, "One Hundred Rules for NASA Project Managers," 9 July 1996. [Online]. Available: https://www.projectsmart.co.uk/white-papers/100-rules-for-nasa-project-managers.pdf. [Accessed 15 December 2018].

[4] Merriam-Webster, "Communications," 2018. [Online]. Available: https://www.merriam-webster.com/dictionary/communications. [Accessed 26 November 2018].

[5] Manager Tools, "Using DISC for Effective Communications," 2018 November 2018. [Online]. Available: managementtools.com.

[6] Available: www.rouloconsulting.com. [Accessed 14 December 2018].

[7] E. J. Roulo, Interviewee, *Aerospace Structures Consultant*. [Interview]. 15-18 September 2018.

CHAPTER 2

Scope Management

Project scope is defined by a set of requirements that characterize the desired products or services and the conditions in which the project team will execute its work.

Scope management involves the set of activities led by the PM to document, baseline, and control the approved scope for the project. Figure 2.1 depicts the recommended steps specific to scope management along with the relationship between scope management and the other PM processes described within this publication.

Scope management is the most important step in managing a project as its outputs form the basis for all other work performed within the project. Project scope management occurs throughout the entire project life cycle as change within a project is common. Changes to the scope can result from a change in the customer's requirements and constraints. Additionally, all project planning in the early stages of a project are based on the team's current understanding of the project scope and any complexities associated with achieving the project's objectives. As the project progresses, the team's understanding of these characteristics will progressively evolve which may result in a need to adjust the project scope.

The goal of scope management is to reduce the occurrence of avoidable change by eliminating ambiguity in the project scope while ensuring all members of the team understand and follow the scope.

> *Document what is agreed upon, deliver what is agreed upon, and do no more.*
> —Eric J. Roulo, Aerospace Structures Consultant [1]

Example project. The project used to demonstrate the project management process within this publication is the fictitious Illudium Q36 Space Modulator depicted in several Warner Brothers cartoons [2]. In this project the customer (the Federation) has identified a need to respond to a new type of threat faced by its forces. Based on an analysis of the organization's mission needs, a project was chartered to develop a solution. The project charter authorizes the use of resources to modify existing systems or build a new system to fill this gap in capability. The project has been contracted to a firm with expertise in weapon systems development. The PM for this project is an employee of the firm awarded the contract. The contract documentation includes a contract, scope of work (SOW), the original charter, and a system performance specification (PSpec) document.

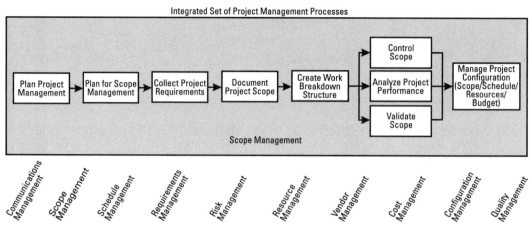

Figure 2.1   The scope management process.

Plan Project Management

Projects are chartered to achieve one or more goals. Project management uses processes, tools, and techniques to achieve objectives in pursuit of these goals. Planning is the thought-based process of identifying the activities and conditions required to achieve a goal or objective.

The overarching plan for the project is the project management plan (PMP) which contains the integrated set of plans to achieve the objectives associated with each PM process. Therefore, each process area described in the following chapters requires a level of planning to achieve its objective in support of the project goal. The PMP serves as the capstone document for all plans.

> *PMBOK Guide–Sixth Edition: Section 4.2 Develop Project Management Plan, provides guidance for defining, preparing, and coordination all planning associated with the project [3].*

The PMP is a living dataset reflecting the current plan for executing the project. The process diagrams reflected within each chapter of this publication call for updating the PMP and any other project artifacts. While the chapters within this publication present each process in a sequential fashion, the processes are executed simultaneously. The results of one process will affect one or more of the other processes and the SMEs recruited to lead each of these processes are supporting members for the other processes.

> *Plans are of little importance, but planning is essential.*
> —Winston Churchill, former British Prime Minister\

The review and evaluation of existing plans are continuous activities within project management. Plans created during the proposal and contracting steps are based on the understanding of the project and conditions at that time. The understanding of the project will grow and change over time and assumptions made in one phase of the project may turn out to be less than accurate. Changes in the

Table 2.1   Roles Associated with Scope Management

Role	Responsibility
Customer lead	The member of the customer's organization charged to manage the project from the customer's perspective. Has final authority over all additions, changes, or deletions to project scope, schedule, and performance.
Project manager	The person accountable for project success, the PM leads, monitors, and supports the scope management processes and procedures.
Procurement officer	Member of the project manager's team that can authorize new agreements and changes to agreements between the project team and the customer or between the project team and its vendors.
Subject matter experts	Members of the PMs organization who are engineers, scientist, and analysts with expertise in one or more area of the project.
Resource managers	Members of the PM's organization that develop and supply technical and project resources to a project team.
Work package lead	The person assigned responsibility and accountability to deliver a work package. May be a member of the PM's organization or the vendor organization if the work package is outsourced.
Project scheduler	Member of the project manager's team that performs the consolidation, structuring, publication, and management of the project schedule. Role may be performed as a secondary function by the PM or by a dedicated resource.
Requirements manager	Member of the project manager's team that performs the consolidation, structuring, publication, and management of requirements. May be performed as a secondary function by the engineers and analysts or by a dedicated resource.
Risk manager	A member of the project team with specialized training and skills in the identification and management of project risk.
Quality lead	A member of the project team with specialized training and skills in establishing processes and practices aimed at delivery products that meet customers' expectations.
Configuration manager	A member of the project team with specialized training and skills in establishing processes and practices required to manage the configuration of items that describe the final product or are a part of the final product.

customer organization, their environment or project constraints will also signify the need to replan.

Throughout the scope management process, information is collected and interpreted to understand the scope and to develop the initial project management plan. As the planning efforts for each of the other PM processes described in this publication are completed, the results are also incorporated into the PMP as subchapters or appendixes.

If a project scope statement was created prior to this step, this information is updated and reused in the PMP. The following information elements should be covered within the PMP along with the detailed plans created for each process area.

Project overview. A short description of the project including the problem or change to be achieved along with the expected outcome.

Project stakeholders. The stakeholder list developed and updated during each of the planning activities.

Product or service description. An explanation of the product or service to be delivered by the project.

Deliverables. A description of each direct and supporting deliverable for the project.

Acceptance criteria. The set of conditions by which the deliverables will be accepted by the customer. This initial list of criteria will be expanded upon during quality management planning. (See Chapter 10)

Source documents. A list of each document used as input to the project planning and scope planning processes. Examples: Contract, Statement of Work, ICD, PSpec, and / or Clarification Memos.

Contributing factors. A description of any potential factors associated with the project. Factors may include urgency, political sensitivity, or implication to the organization if the project does not deliver or unique operational conditions expected of the solution.

Contractual conditions. The type of contract used for this project along with any contractual conditions for surveillance, reviews, and reporting. Also included here are any penalties or incentives associated with the contract. An overview of contracting types is provided in Chapter 7.

Financial conditions. A description of the type and source of the funding to be used for this project. The funding source for the project may be contingent on an external condition or funds may be available in increments over several budget years. The project schedule, budget, and resource plans must reflect when funding will be available. Projects supporting US Government customers are funded with various types of money. Each type of money has limitations as to how it can be used and the timeframe in which the funding can be spent. Example funding types include procurement, operations and maintenance, and research and development.

Cost management. A definition of the cost management techniques such as earned value management required or planned for the project. The process and methods for performing cost management will be defined during cost management planning (see Chapter 8).

Assumptions and clarifications. A list of all assumptions made by the project team in developing the project management plan along with any points that required clarification during these processes.

Project boundary and exclusions. A description of items explicitly excluded from the project scope. A graphic depicting the organization or system and the demarcation points to external entities is a good way to help the reader to visualize the project boundary.

Major milestones. A list of the key dates or milestones defined within the contract along with the milestones that when reached allow for invoicing to the customer. These initial milestones will be expanded upon during schedule management planning (see Chapter 3).

Phase gate reviews. Phase gate reviews are placed at critical periods in the project life cycle where work performed up to a given point drives the work downstream in the project. Phase gate reviews require planning, documentation development, and time to perform the review (see Chapter 3).

Resource requirements. High-level description of the human, facility, and equipment requirements. This initial description will be expanded upon during resource management planning (see Chapter 6).

Life cycle support. A description of any contractually mandated operations support, spare parts inventories, warranties, or end user training.

Initial risk assessment. Risks identified during the project proposal, negotiation, and planning processes. This initial list will be expanded upon during the risk identification steps within the risk management process (see Chapter 5).

Quality management strategy. A description of any quality management requirements mandated by the contract. The processes and methods for executing quality management will be defined during quality management planning (see Chapter 10).

Plan Scope Management

Understand project environment. The PM starts scope management by documenting the environment in which the project will exist. Projects are executed within the PM's organization which has structure, roles, responsibilities, and policies associated with execution of project activities. The project customer is also an organization with a structure, roles, responsibilities, and policies. The customer organization also resides within an environment that includes factors such as the intended users, ecosystem, and potentially impacted parties within the ecosystem. These factors are likely to have a set of policies and regulations that become scope conditions. The intent of this step is to establish an understanding of the external factors that may influence the planning process.

Collect project documents. How the project was authorized by the customer will define the level of information available on the scope of the project. The information available at the beginning of internal projects may consist of little more than the business case data presented to obtain approval for the project. This potential lack of clarity puts the onus on the PM to document the scope for the customer. The information provided for projects contracted to the PM's organization should include detailed requirements, scope documents, and contractual clauses. However, more documentation does not necessarily equate to scope clarity. This abundance of information can be filled with implied requirements, subjective performance criteria, and contradictory statements. In these cases, the PM must analyze all the documents to identify all scope information for the project. The goal here is to identify all available and potentially missing information related to the project scope.

Get help. It is never too early to engage parties with subject matter expertise (SME) in each area of the project. SMEs exist within the PM's organization as functional managers, experienced project managers, and those who have led similar projects. Technical and regulatory stakeholders may also provide interpretation for specific requirements or contract clauses. Engaging these parties early serves to build relationships that are critical to project success. Examples of these relationships are provided in later chapters.

Identify stakeholders. A stakeholder is any party with an interest in the outcome of the project. Stakeholders can have an implicit interest as the sponsors or users of the project outcomes or they can have implied interest driven by social, political, or economic motivations. This step involves developing an initial project stakeholder list that identifies all the parties with an interest in the outcome of the project. This initial list may include groups of parties rather than specific people.

The motivation for interest and degree to which each stakeholder group could impose requirements on the project is also documented at this time.

> *Technical and regulatory stakeholders are those with explicit or implied authority to impose requirements on projects. Technical and regulatory authority is assigned to independent representatives of the customer organization or jurisdictions in which the project resides. Technical and regulatory authorities are charged with the responsibility of enforcing standards for safety, interoperability, quality, security or processes.*

Document roles. In order to manage the project scope, the PM requires an understanding of the roles and authorities associated with the scope. For each activity within scope management, the PM identifies the parties with a role in the process. A good way to characterize these roles is by responsible, accountable, consulted, or informed (RACI).

Responsible. The parties who will do the work. Every task in the list will have at least one responsible party.

Accountable. The party charged to ensure the work is completed. In matters involving scope, the accountable party has the authority to approve the initial scope baseline and any changes to the scope.

Consulted. Those whose input is required for the task. SMEs and technical or regulatory stakeholders fall into this group.

Informed. The parties who should be kept informed of the task and its outcome.

The result of this process is a responsibility assignment matrix (RAM) or RACI chart. Table 2.2 provides a simplified example of a RAM where the tasks make up the first column and the remaining columns define each party's role.

It is important to notice in this example that the leads for the other process areas have roles in the tasks that involve scope management. This example highlights the need for the PM to maintain common awareness and collaboration across all leadership within the project team.

Define processes. The project team identifies the processes required for developing the scope documentation, achieving scope approval, and authorizing changes to the scope. The data used to define the scope and decisions made during this process constitute items that require configuration management. The processes

Table 2.2   Scope Management RACI Matrix

Task	Project Sponsor	Customer	PM's Leadership	Project Manager	Resource Providers	QC Lead	Scheduler	Risk Manager
Collect Req	C	C	I	A	I	R	R	R
Approve Req	A	R	R	R	C	C	C	C
Develop WBS			I	A	I	R	R	R
Define Scope	C	C	I	A	I	R	R	R
Approve Scope	A	R	R	R	I	I	I	I
Control Scope	C	C	C	A	C	C	C	C
Change Scope	A	R	R	R	C	C	C	C

used to manage the configuration of documents, data, and products are described within Chapter 9. In this planning step, the PM works with the configuration manager to identify the scope data, documents, and decisions to be managed under the project's configuration management processes.

Document project life cycle. A project passes through a natural progression or life cycle starting with the initial definition of the project and potential solutions through to development, delivery, and sustainment of the desired solution. Project life cycles are divided into phases that reflect the maturation of the selected solution as shown in Figure 2.2. Each of these phases have a set of expectations for measuring the maturity of the solution's development which are used to review the project before it can move to the next phase.

The specific project life cycle used on projects is often prescribed by the customer's organization and its organizational policies. In the absence of a prescribed life cycle, the PM relies on input from his or her SMEs and the organizational process assets (OPA) to identify the appropriate model [3].

Products eventually reach a point of usefulness which could spawn a new project to replace the obsolete item which is then decommissioned. Projects that support government and military organizations almost always have a portion of the life cycle that addresses how the product will be decommissioned and disposed. Recognizing this eventuality helps to ensure the product design and development addresses the disposal of items that can be harmful to people and the environment.

Life cycle approaches. How a product is developed may follow one of several approaches or models. Where the life cycle in Figure 2.2 provides the benchmark for the maturation of the product, the approach selected defines how the full set of requirements will be satisfied. Life cycle approaches allow the customer to achieve the solution all at once or in portions over time.

The decision to build the product in increments over time can be based on availability of funds, availability of technology, or the need to see immediate results

Project Phase/Key Milestone						
Pre -Phase A	Phase A	Phase B	Phase C	Phase D	Phase E	Phase F
Concept Design	Concept and Technology Development	Preliminary Design & Technology Completion	Final Design and Fabrication	System Assembly, Integration & Test Launch	Operation & Sustainment	Decommissioning
◆	◆ ◆	◆	◆	◆	◆	◆
CoDR	SRR SDR	PDR	CDR	TRR FRR	PRR	DR

CoDR: Concept Development Review
SRR: System Requirements Review
SDR: System Development Review
PDR: Preliminary Design Review
CDR: Critical Design Review
TRR: Technical Readiness Review
FRR: Functional Readiness Review
PRR: Production Readiness Review
DR: Decommissioning Review

Figure 2.2 Common technical project life cycle.

in priority areas. In other cases, the customer may not have a clear understanding of what constitutes a satisfactory solution. In these cases, the approach may be to build a little, test a little, learn a little over several cycles. The results of each cycle are used to define the focus areas of the next spiral. All these approaches align in some fashion to the generic life cycle. The alignment may come in the form of repeated development phases or with the full life cycle repeated for each increment. The life cycle approach may be defined or implied within the project documentation. The PM relies on input from his or her SMEs to identify the directed or implied approach and then acquires concurrence from those with authority to accept or change the project scope [3].

Collect Project Requirements

A project is a temporary effort intended to achieve one or more goals. Projects have a life cycle with a starting point and a defined end. Projects create deliverables in the form of products or services that either represent the project's goal or support achievement of the project's goal. A project scope is defined by a set of requirements that characterizes the desired products and services (the goal) along with the conditions in which the project team will execute its work.

At the most basic level, a requirement is something that is needed or is compulsory. Requirements typically start out in the form of a business or mission need (business requirements) which is the reason for the project. Product requirements describe the characteristics of the product or service to be developed by the project and project requirements describe the conditions by which the new product or services is to be delivered. Project requirements (the conditions) include major milestones, contract clauses, data to be delivered, and cost or schedule constraints. At this point in the project life cycle, product requirements are likely to be a very high-level description of the functionality the product or service is to perform. The project scope is the composition of all business, product, and project requirements as shown in Figure 2.3.

These requirements do not always come wrapped up in a clear and concise list. In this step the goal is to establish a single list of all requirements defined in the project initiating documents that will serve as the baseline set of requirements for the project.

These documents may come in the form of

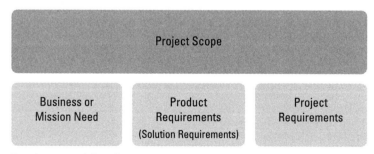

Figure 2.3   Requirements define the scope.

- Business or mission need statements;
- Initial capability descriptions;
- PSpec;
- Business case documents;
- Project charter;
- Standard contract clauses;
- Scope statements;
- Contract documents.

The U.S. Federal Acquisition Regulation (FAR) establishes the rules and guidelines followed by United States Government agencies when acquiring goods, products, and services. The FAR contains hundreds of standard clauses and provisions that are applied to government contracts. FAR clauses are intended to protect the rights of both the government and the contractor [4]. These standard clauses constitute project requirement statements.

Statements within the collected documents are not likely to be labeled as requirements. The project manager uses subject matter experts (SMEs) in contracting, requirements management, engineering, quality, risk, and other specialties to identify and compile the identified requirement statements into a single list. Typically, project technical lead is responsible for compiling product requirements (PSpec), project manager for compiling project requirements, and quality for compiling user acceptance requirements (Standard Clauses, etc.) Requirements are logged in a tabular or relational database so that attributes can be added to each statement. Each requirement statement is linked to its source document to maintain traceability to its source.

The documents provided with the example project (Q36 Project) included the formal contract, the statement of work, the system PSpec document, and a series of contract clauses attached to the contract. In this example, the PM and his team

Number	Name	Description	Source	Status
1000000	Q36 Illudium Space Modulator Project	Contract 2017.09.08-5	Project Contract	
2000000	⊟ Q36 Space Modulator Requirements	Project to deliver new or revised systems that will enable the commander to respond to new threats	Project SOW and Charter	Draft
2010000	⊟ Problem Statement	New Threats Require an Improved Capability	Project SOW and Charter	Draft
2010100	⊟ Sense Targets	Ability to Sense Targets	System Performance Specification	Draft
2010110	⊞ Sense Range	The system shall sense targets within 100 miles	System Performance Specification	Draft
2010200	⊞ Destroy Targets	Ability to Destroy Targets	System Performance Specification	Needs Specification
2010300	⊞ Track Targets	Ability to Track Targets	System Performance Specification	Needs Specification
2010400	⊞ Identify Targets	Ability to Identify targets	System Performance Specification	Needs Specification
2010500	⊞ Target Discrimination	Ability to discriminate between enemy and friendly targets	System Performance Specification	Draft
3000000	⊟ Project Requirements		Project SOW and Charter	Draft
3010000	Milestone Reviews	The contractor will support milestone reviews	Project SOW and Charter	Draft
3020000	Incremental Delivery	The project will deliver the project over 4 discrete increments	Project SOW and Charter	Draft
3030000	Data Requirements List	Contract Data Requirements List	Project SOW and Charter	Draft
3040000	⊞ Earned Value Management System	Contractor will provide earned value management data for the project	Project SOW and Charter	Needs Specification
3050000	⊞ Quality Management System	The contractor will employ a quality management system	Project SOW and Charter	Needs Specification
3060000	Configuration Management	The contractor will utilize a set of configuration management processes and data	Project SOW and Charter	Draft
3070000	Requirements Engineering	The contractor will provide requirements engineering	Project SOW and Charter	Draft
3080000	Systems Engineering	The contractor will provide systems engineers, designers and analysts as required	Project SOW and Charter	Draft
3090000	Systems Integration	The contractor will integrate all sub-systems and components into a single solution	Project SOW and Charter	Draft
3100000	Systems Prototyping	The contractor will develop a functional prototype for each major subsystem prior to proceeding to design	Project SOW and Charter	Draft

Figure 2.4 Initial requirement matrix.

of SMEs have identified a series of requirements statements within the provided documents and compiled them into the initial requirements matrix in Figure 2.4.

Notice several of the requirements are labeled as needs specification to denote ambiguity in the current statement. For example; the statements for identifying, tracking, and destroying targets are missing key performance parameters. The project manager works with the project technical lead, customer, leadership, and the SMEs to remove all ambiguity from the requirements list. The project team's ability to meet the customer's expectation is contingent on the accuracy of these requirements. The intent of this task is to have a clear description of the product or service to be delivered in terms understood by the customer that will serve as the objective measurement for project success.

Figure 2.5 reflects the clarifications added to the requirements based on negotiations with the customer. In some cases, the clarifying information was found within the contract documents and in a few cases, the customer was required to submit a memo clarifying the requirement. Lastly, all requirements were noted as approved after the accountable party agreed to the final set.

The product requirements will be used as input to the requirements management processes described later in this publication. Throughout the project life cycle these requirements will be progressively elaborated and refined to support the design and delivery of the solution.

Document Project Scope

Up to this point, the information regarding the project scope is found in multiple locations. Scope information was first found in the customer provided documentation in which product and project requirements were compiled into the requirements matrix. This information was used to clarify any potential gaps or conflicts between

Number	Name	Description	Source	Status
1000000	Q36 Illudium Space Modulator Project	Contract 2017.09.08-5	Project Contract	
2000000	Q36 Space Modulator Requirements	Project to deliver new or revised systems that will enable the commander to respond to new threats	Project SOW and Charter	Approved
2010000	Problem Statement	New Threats Require an Improved Capability	Project SOW and Charter	Approved
2010100	Sense Targets	Ability to Sense Targets	Initial Capability Description	Approved
2010110	Sense Range	The system shall sense targets within 100 miles	Initial Capability Description	Approved
2010111	Range of 100 miles	The system shall sense a target at a distance from the sensor up to 100 miles	Initial Capability Description	Approved
2010112	Target Environment	The system shall sense a target in space atmosphere as defined in Attachment C of the ICD	Initial Capability Description	Approved
2010113	Target Sense Speed	The system shall sense targets moving at speeds between 1 mile and 900 miles per hour	Initial Capability Description	Approved
2010200	Destroy Targets	Ability to Destroy Targets	Initial Capability Description	Approved
2010210	Engagement Range	The system shall engage target at a range of 75 miles.	Initial Capability Description	Approved
2010220	Engagement Speed	The system shall traverse distance between shooter and target before target can reach shooter	Initial Capability Description	Approved
2010300	Track Targets	Ability to Track Targets	Initial Capability Description	Approved
2010310	Target Tacking	The tracking system shall maintain a track of target movements	Initial Capability Description	Approved
2010311	Tracking Speed	The system shall track targets moving up to 900 miles /hours	Initial Capability Description	Approved
2010400	Identify Targets	Ability to identify targets	Initial Capability Description	Approved
2010410	Target Identification	The system shall match the target against a set of known targets	Initial Capability Description	Approved
2010500	Target Discrimination	Ability to discriminate between enemy and friendly targets	Initial Capability Description	Approved
2010510	Designate enemy targets.	The system shall designate enemy targets on the shooters display	Initial Capability Description	Approved
2010520	Designate friendly targets.	The system shall designate friendly targets on the shooters display	Initial Capability Description	Approved
3000000	Project Requirements		Project SOW and Charter	Approved
3010000	Milestone Reviews	The contractor will support milestone reviews	Project SOW and Charter	Approved
3020000	Incremental Delivery	The project will deliver the project over 4 discrete increments	Project SOW and Charter	Approved
3030000	Data Requirements List	Contract Data Requirements List	Project SOW and Charter	Approved
3040000	Earned Value Management System	Contractor will provide earned value management data for the project	Project SOW and Charter	Approved
3040100	EVMS Reporting	The contractor EVMS will report BCWS, BCWP and ACWP for each work package in the WBS	Customer Clarification Memo	Approved
3040200	EVMS Indicators	The contractor EVMS will report SPI, CPI and TCPI performance indexes for each work package in the WBS	Customer Clarification Memo	Approved
3050000	Quality Management System	The contractor will employ a quality management system	Project SOW and Charter	Approved
3050100	ISO 9001:2015	The contractor's quality management system will be certified compliant with the requirements defined in ISO :9001:2015	Customer Clarification Memo	Approved
3060000	Configuration Management	The contractor will utilize a set of configuration management processes and data	Project SOW and Charter	Approved
3070000	Requirements Engineering	The contractor will provide requirements engineering	Project SOW and Charter	Approved
3080000	Systems Engineering	The contractor will provide systems engineers, designers and analysts as required	Project SOW and Charter	Approved
3090000	Systems Integration	The contractor will integrate all sub-systems and components into a single solution	Project SOW and Charter	Approved
3100000	Systems Prototyping	The contractor will develop a functional prototype for each major subsystem prior to proceeding to design	Project SOW and Charter	Approved

Figure 2.5 Approved requirements list.

these statements. Throughout this process, clarifying information was provided by the customer, decisions were made by the accountable party, and in other areas assumptions were made by the project team. The objective of this step is to coalesce all scope information into a single document. The scope information forms the foundational information within the project management plan (PMP).

The need to create a stand-alone document containing the project scope occurs frequently when the project is internal to the PM's organization. External projects typically result in a formal contract or agreement between the customer and the contractor, but internal projects often lack this formality. The Project Scope Statement is an effective means to achieve concurrence on the project scope and pertinent overarching information. The format for this document varies by organization and project size but the following key topics should be covered within a project scope statement [3].

Project overview. A short description of the project including the problem or change to be achieved along with the expected outcome.

Product or service description. An explanation of the product or service to be delivered by the project.

Deliverables. A description of each direct and supporting deliverable for the project.

Acceptance criteria. The set of conditions by which the deliverables will be accepted by the customer. This initial list of criteria will be expanded upon during quality management planning.

Project boundary and exclusions. A description of items explicitly excluded from the project scope. A graphic depicting the organization or system and the demarcation points to external entities is a good way to help the reader to visualize the project boundary.

Create WBS

The work breakdown structure (WBS) is a hierarchical representation of the project deliverables. Like an organization chart where the CEO is at the top of the chart, the top level of a WBS is the end item product or service to be delivered by the project. The next level below the CEO is typically the senior executives that lead segments of the business. In the WBS the next layer is the set of major building blocks for the product or service. The building blocks may represent major functionality or structural elements of the product. For example; the major structural elements of an aircraft project could include the fuselage, wing span, power plant, and control systems. Each building block is progressively decomposed to smaller and smaller discrete blocks. The goal here is to break the work down to a point in which each block can be assigned to a single responsible party. Figure 2.6 represents the initial WBS for the Q36 based on the project team's current understanding of the project.

In Figure 2.6 the top level of this WBS represents the Q36 which is the end item product to be delivered by the project. The next layer down represents the major elements of the weapon system described within the Product Requirements. The elements under 101000 Project Management reflect the deliverables required within the project contract and scope of work. These items reflect the Project Requirements. Each of the level-two building blocks are further decomposed by aligning

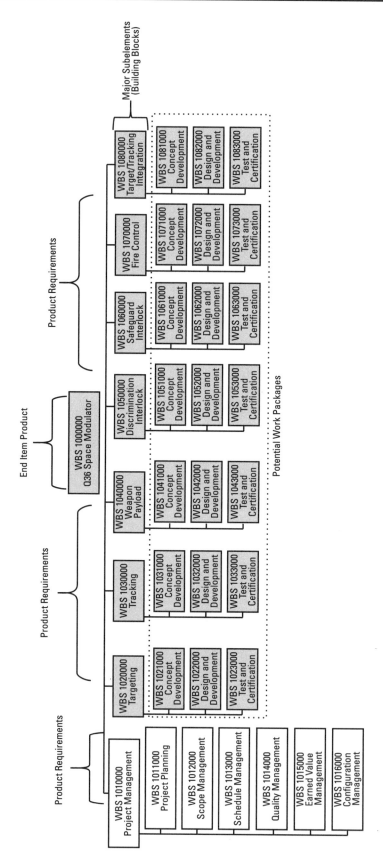

Figure 2.6   Example WBS.

the requirement statements under the respective building block. The goal to this step is to create a visualization of the project scope. This visualization supports ratification of the scope with the customer and communicating the scope to the project team. The initial WBS created during this step is expanded upon and used in many of the subsequent project management processes.

Validate Scope

Scope validation involves the activities associated with achieving acceptance for each deliverable. The objective of this activity is to ensure that each deliverable is evaluated and approved using predefined, objective criteria. Chapter 10 describes how the quality lead helps to document these acceptance criteria. These criteria are used to validate each deliverable prior to its release to the customer.

Lack of clear definition of the customer's acceptance criteria in the project contract can adversely impact the project's success and customer satisfaction. These acceptance criteria are project requirements and must be documented in the project scope, contract, and planning documents.

As each deliverable is prepared for release to the customer, quality lead, the technical lead, and the PM validate the deliverable to ensure it meets the requirements for the deliverable.

A rejected deliverable is one that has been deemed by the customer as noncompliant with their expectations. At this point, the project team must remedy the situation and then resubmit the deliverable to the customer.

Failure to meet expectations may be due to the project team's inability to meet a requirement as defined within the current scope of work. In some cases, the project team requires more time to resolve the temporary noncompliance issue and in other cases the team may not anticipate being able to meet the requirement within the current cost and schedule constraints.

For temporary issues, the project team may request a variance to deliver a product until the requirement can be met. The resolution to the more drastic issues will likely require either a permanent waiver for the requirement or an increase in cost and schedule to achieve the desired result. Either of these conditions should indicate a need for modification to the project scope and supporting documents which should be managed using the change management processes defined during the configuration management planning activities (see Chapter 9).

The disposition of each deliverable's acceptance, rejection, or change conditions is tracked within the project's quality management data set explained in Chapter 10. This data is used by the PM to monitor project performance and to identify potential issues.

Control Scope

Controlling the project scope involves the activities associated with monitoring the performance of work and ensuring that work products satisfy the scope while identifying any potential changes to the scope.

Validating work against the scope and identifying potential changes to scope requires management of the scope documentation and data. Upon completion of the scope and project management planning activities, the project scope, PMP, requirements, and WBS are approved and set to a baseline. This baseline of information is the benchmark used for all project monitoring and control activities. Any approved changes to the scope, PMP, requirements, or WBS are saved to the current baseline or to a new baseline either of which become the new benchmark for project monitoring and control.

Monitor performance. Undocumented changes to the project scope can occur when a member of the project team agrees to do work that is not within the project scope and requirements. These changes may be the result of misinterpretation of the scope by the team member, a casual request from the customer, or any number of external conditions that result in delays or additional resources. Often the only way to catch these changes is when project performance indicators do not match the project baselines. The PM uses data collected during each of the PM processes to measure the team's performance. The quality management data previously discussed is an example of this type of data.

If you can't measure it, you can't manage it.
—Peter Drucker

The following is an example how the PM can detect an undocumented change: The concept development work package for Q36 Targeting system is contracted to a vendor. The vendor's contract has a set budget and expected completion date. The scheduled delivery date is drawing near, and the vendor's invoices reflect that most of the allocated funding has been spent. However, the vendor reports that they are only 60% complete on the task. This misalignment between the baseline plan and the actual indicates a potential issue. When the vendor is asked to explain the performance issue, he states he had to redo his design twice to support new requirements communicated during project reviews. As it turns out, the new requirements were communicated by the PM's chief engineer as suggested improvements to the design. The vendor took these suggestions as direction and then acted upon them. The potential implications of adding new requirements are far greater than additional cost and time for this vendor. These changes could impact other system designs downstream from this design, further impacting the project's performance. Monitoring the vendor's performance data allowed the PM to detect this change before it could cause further damage to the project.

Manage change conditions. Any change to the product or project requirements will impact the project's cost, schedule, or performance. Therefore, the approval of any change can come only after the change has been thoroughly assessed for all impacts. Changes to the scope are managed through a set of processes and business rules designed to ensure that all approved changes are necessary, and the implications of the changes are fully understood.

Change management is a function of the configuration management processes which are described in Chapter 9. The configuration management lead recruited to the project helps the PM to define the processes required for evaluating, approving, and managing all change conditions. CM planning involves the documentation of the standards, policies, practices, procedures, and instructions that guide the project team in managing the configuration of its final deliverables as well as the intermediate information, mechanisms, and assemblies that lead to final products. The project scope as reflected in the Contract, SOW, PSpec, PMP, and the requirements matrix fall into this category of items requiring configuration management.

Summary

Scope management is the first and most important step in managing a project as its outputs form the basis for all other work performed within the project. Project scope management occurs throughout the entire project life cycle as changes within projects are common.

Projects are chartered to achieve one or more goals. Project management uses processes, tools, and techniques to achieve objectives in pursuit of these goals. Planning is the thought-based process of identifying the activities and conditions required to achieve a goal or objective.

The overarching plan for the project is the PMP that contains the integrated set of plans to achieve the objectives associated with each process. The PMP serves as the capstone document for all plans developed within the processes described in each chapter of this publication.

Product requirements characterize the product or service to be developed by the project. Project requirements describe the conditions by which the new product or service is to be delivered. Project requirements (the conditions) include major milestones, contract clauses, data to be delivered, and cost or schedule constraints.

Product and project requirements do not always come wrapped up in a clear and concise list. The PM along with the project SMEs creates a single list of all requirements defined in the project initiating documents. The goal of collecting requirements at this stage in the project is to establish a single set of clear and concise parameters by which project success is to be measured.

The full set of information that defines the project scope is often spread throughout several documents such as the contract, contract clauses, scope of work, and other supporting documents. The PM coalesces all this information into a single scope document.

The top level of a WBS is the end item product or service to be delivered by the project. The next level below in the WBS is the set of major building blocks for the product or service. The WBS allows the PM to break the work down to a point in which each block can be assigned to a single responsible party.

Any change to the product or project requirements will impact the project's cost, schedule or performance. Therefore, the approval of any change can come only after the change has been thoroughly assessed for all impacts. Changes to the scope are managed through a set of processes and business rules designed to ensure that all approved changes are necessary and the implications of the changes are fully understood.

References

[1]   Roulo, E. J., Interviewee, *Aerospace Structures Consultant* [interview], September15–18, 2018.

[2]   Jones, C. M., Director, *Haredevil Hare* [film], Warner Brothers Cartoons and Vitaphone, 1948.

[3]   *A Guide to the Project Management Body of Knowledge: PMBOK Guide*, Sixth Edition, Newtown Square, PA: Project Management Institute, Inc., 2017.

[4]   US GSA, US DoD, NASA, Federal Acquisition Regulation, Washington, DC: U.S. General Services Administration, US Department of Defense, National Aeronautics and Space Administration, 2019.

CHAPTER 3

Schedule Management

A project schedule represents the plan to deliver the products and services associated with a chartered work effort. A project schedule includes for each element of work: a description, duration, start and end dates, resources allocated, and dependencies with other elements of the schedule. Schedule management is the process led by the PM to define, communicate, monitor, and control the project schedule. Project schedules and schedule management apply to all projects regardless of the adopted life cycle approach (sequential, iterative, incremental, agile). The objective of schedule management is to lead the timely completion of a project. This chapter breaks down the sometimes-arduous process of creating and managing a schedule into a series of straightforward steps.

> *The purpose of schedule management is to provide the framework for time-phasing, resource planning, coordination, and communicating the necessary tasks within a work effort.*
> —NASA Schedule Management Handbook -2010 [1]

There are several sources of guidance available for schedule planning and management. The processes and techniques offered in this chapter are a generic representation of common steps found in these sources.

Figure 3.1 depicts the recommended seven-step process for schedule management. The best way to think about scheduling is to answer the question: *Who* should do w*hat* and *when* does it need to be done? The activities in the first step involve identifying what needs to be done.

The subsequent steps in the process involve assigning attributes to each activity. The first set of attributes define *when* (*sequencing, duration,* start and end dates, and dependencies) the activities are to be performed. Next, the who is defined by the *resource* attributes that include people but may also include facilities, equipment, and support systems. The schedule is an amalgamation of these data points that can be presented in tables, bar charts, or calendars. As a living data set, the schedule must be continually controlled, monitored, refined, and updated as the project plan materializes.

Figure 3.2 is an example of a project schedule that results from performing these steps.

The schedule management process is performed throughout the entire project life cycle. The project schedule is a living data set used by all stakeholders. As depicted in Figure 3.3, the project schedule starts off as a high-level view of

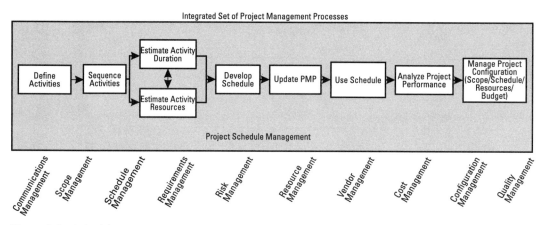

Figure 3.1   Schedule management process.

the project and is progressively elaborated as the project progresses. One should not confuse progressive elaboration with unmanaged schedule change. The start, end, and durations for the top-level elements of the schedule remain intact; what changes is the level of detail under each element as the project team develops its understanding of the work required to complete the top-level elements.

Unfortunately, the PM is not always afforded the desired time and resources required to complete the project. In many cases, the project manager is tasked to lead a project with a sooner-than-desirable due date and inadequate resources to meet this due date. The project schedule is the result of negotiations between the PM, the customer, the PM's leadership, and resource providers to achieve the most acceptable plan to deliver the project balanced between scope, cost, and schedule.

Figure 3.4 depicts the forces at play in negotiating an acceptable project schedule. The scale of each of these forces is intentionally shown disproportionate to the resulting schedule to amplify the effect these forces have on the schedule development. In real-world projects, these factors never match what the project manager desires. For example:

- Customers often want more than they can afford *(what)*;
- Customers often need the project completed faster than reasonable *(when)*;
- Resource providers rarely have their expert people available when needed *(who)*.

This diagram also helps to demonstrate that changes to one item directly affects the other two items. Adding more scope impacts the time and resources required. Conversely, adding more resources may not reduce the time required to complete a set of tasks. Doubling the number of people or extending working hours on a single set of tasks can result in confusion, duplicate work, work area congestion, mistakes, and other inefficiencies. Assigning 10 painters to paint a 10-foot by 10-foot room will surely result in one thing: eight painters watching the other two work.

Schedule management is the process led by the PM to define, communicate, monitor, and control the project schedule. A project schedule represents the plan to deliver the products and services associated with a chartered work effort. A project

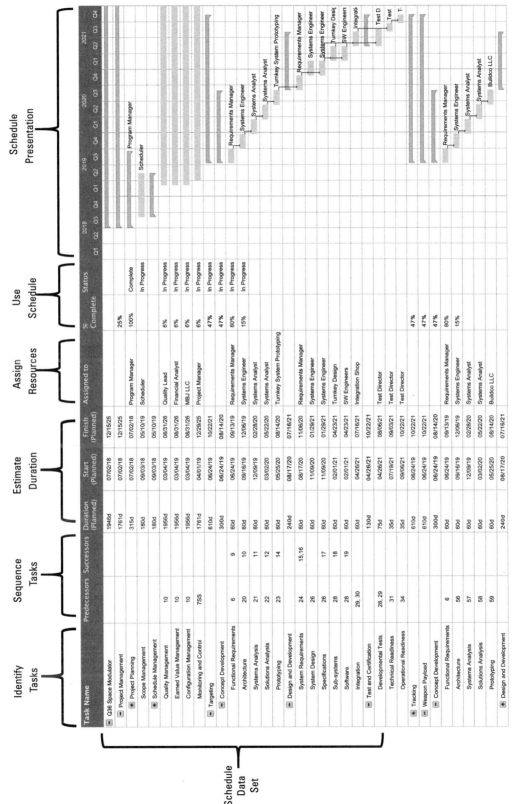

Figure 3.2 Results of the schedule management process.

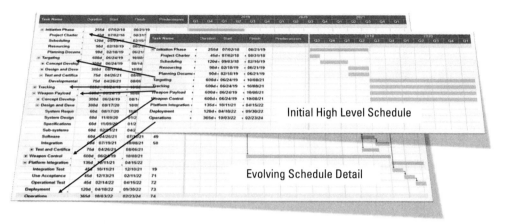

Figure 3.3   Evolving schedule detail.

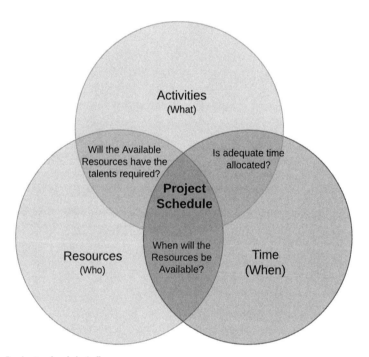

Figure 3.4   Project schedule influences.

schedule includes for each element of work: a description, duration, start and end dates, resources allocated, and dependencies with other elements of the schedule. Project schedule management requires a clear understanding of the work, a deliberate planning process, and the ability to negotiate between competing interests to create and manage a schedule that leads to the timely completion of the project (see Table 3.1).

Predominant sources for Scheduling Guidance include: the NASA Schedule Management Guidelines and the Project Management Institute's (PMI) Project Schedule Management Process.

Table 3.1   Roles Associated with Schedule Management

Role	Responsibility
Customer lead	The member of customer's organization charged to manage the project from the customer's perspective. Has final authority over all additions, changes, or deletions to project scope, schedule, and performance.
Project manager	The person accountable for project success, the PM leads, monitors, and supports the vendor management processes and procedures.
Procurement officer	Member of the project manager's team that can authorize new agreements and changes to agreements between the project team and the customer or between the project team and its vendors.
Subject matter experts	Members of the PMs organization who are engineers, scientist, and analysts with expertise in one or more areas of the project.
Functional managers	Members of the PM's organization that develop and supply technical and project resources to a project team.
Work package lead	The person assigned responsibility and accountability to deliver a work package. May be a member of the PM's organization or the vendor organization if the work package is outsourced.
Project scheduler	Member of the project manager's team that performs the consolidation, structuring, publication, and management of the project schedule. Role may be performed by the PM as a secondary function or by a dedicated resource.

Define Activities

Chapter 2 introduced the process for documenting, elaborating, and managing the project scope. During scope management, all currently available information (statement of work, requirements documents, contracts) are analyzed and decomposed to increase the clarity and fidelity of the project scope.

The goal of defining the activities and tasks is to translate the project requirements, as described in the project scope, into a description of the work required to satisfy these requirements.

Work Breakdown Structure

The project requirements define *what* is to be delivered, *how* many are required, and *when* these items are to be delivered. Project requirements form the foundation for the project. The project requirements are organized into a WBS that reflects the major building blocks of the project and any relationships between these blocks. At a minimum, the initial WBS drills down to a point where work packages can be assigned to individual resource pools. These resource pools may be internal departments within the project organization or externally contracted suppliers. Figure 3.6 is an example of a hierarchical depiction of a WBS and the lower-level blocks representing the proposed work packages for the project.

A work package is any unit of work that can be assigned to an individual resource group.

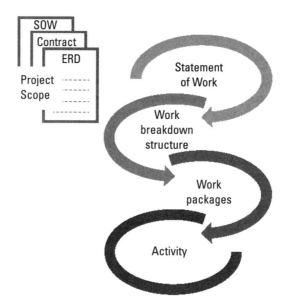

Figure 3.5   Activity definition.

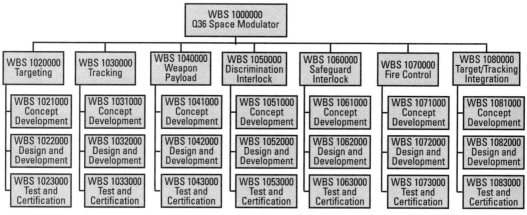

Figure 3.6   Initial WBS.

Work Package Planning

Work packages owners complete work packages by performing activities. The work package owner defines the activities required to deliver each work package. The dates, durations, and sequencing of these activities become the details of the project schedule. PMs should not expect to know everything in the first round of schedule development. In fact, in the early stage of a project the amount of information unknown may be greater than the information known. The initial project WBS, work package list and project milestones are used to communicate talent requirements to the resource providers. These very talents are required to define the project activities and flush out the entire schedule.

The PM is not expected to be an expert on how every task within the project is completed. Work packages allow the PM to establish a demarcation point between the overall project structure and the work performed by specific resources. The PM and the broader project stakeholders are concerned with the delivery of work products and progress toward the delivery of same. The work package owners are primarily focused on the activities performed to deliver their work package and any external dependencies that may affect their work.

The PM must strike a balance between telling the work package owner how to do their work and having no visibility into the work performed within a work package. Adopting a standard set of top-level activities for all work packages is a good tactic to strike this balance. For example, a system engineering effort typically follows a system development life cycle (SDLC) that includes concept exploration, design, system development, testing, integration, deployment, and operations. Other types of project have similar life cycles. By requiring each work package owner to organize their work by a common set of activities, the PM has a point of reference for monitoring the work packages' progress while leaving the details on how these activities are achieved up to the expert performing the work. During this step, the PM works with the work package owners to establish the top-level activities that will provide the required insight and then facilitates integrating the plans for the individual work packages into an integrated master activity list. (Recruiting and selecting work package owners is described in Chapter 6.)

One must be very careful in discerning between adding fidelity to the plan and adding new work to the project. A good rule of thumb is the addition of new work packages or top-level WBS elements typically means new scope. Adding details to items under the existing work package should result in an improved definition of its higher-level WBS element not a new definition of the WBS element.

The top-level of the WBS is the decomposition of *what* the project is to deliver according the approved scope. New details added below each WBS element describe *how* the top-level element will be achieved. The activity list in Figure 3.8 includes the agreed-upon major milestones as well as subtasks defined by the discrimination interlock work package owner.

Indirect Efforts

One may notice the sample WBS in Figure 3.7 does not include the project management efforts. Yet, these efforts do appear in the sample schedules provided in Figure 3.2. This discrepancy was not a mistake on part of the PM, it is an example of an area where there are varying opinions on the purpose of the WBS. Some teach that the WBS contains just the work associated with the end item product (the Q36 SM). Others teach that the WBS should contain indirect efforts required to support the direct work. For example: the U.S. Navy Acquisition Community has adopted a Global WBS (GWBS) that includes direct end item products and indirect support such as engineering, project management, and training. For the sake of this discussion, the latter method is followed here to establish a single list of all work and the sample WBS now reflects indirect support activities.

Regardless of the method followed, the indirect activities required to complete the project must be identified, resourced, scheduled, and coordinated with the direct efforts.

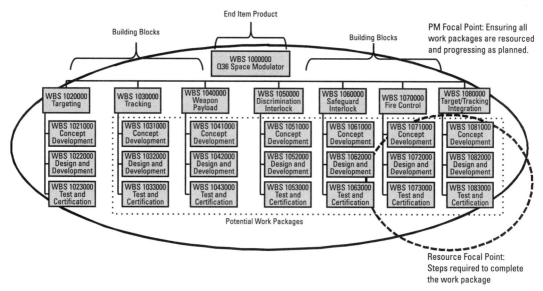

Figure 3.7   Differing focal points.

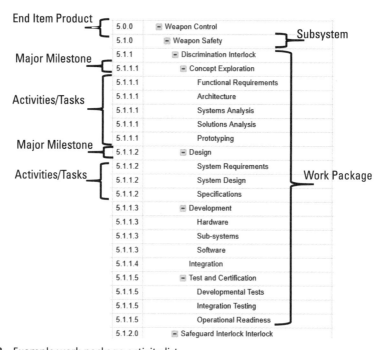

Figure 3.8   Example work package activity list.

Phase Gate Reviews

Phase gate reviews are an important tool for the PM. Yes, reviews can be a distraction and they divert resources away from project execution. However, reviews are often the only opportunity in which the project team can obtain feedback and/or concurrence from the customer. Phase gate reviews are placed at critical periods in the project life cycle where work performed up to a given point drives the work downstream in the project. The goal of gate reviews is to avoid changes to the work products that drive downstream work after the downstream work has commenced. These reviews involve the evaluation of project deliverables such as requirements, specifications, analysis results, testing reports, and designs artifacts. Figure 2.2 depicts the phase gate reviews often required in large acquisition programs.

During each review, a set of work products is reviewed by the project stakeholders to verify that the work product is consistent with the project requirements, standards, and other regulations. In this example life cycle, the concept design is presented at the concept design review (CoDR). At this point in the project life cycle, the project team is recommending a solution concept that will drive the rest of the project. One can imagine the impact on the project if the team was ready to fabricate the fully designed system (phase C) based on the concept design (prephase A) to then learn that the concept design did not include a major safety regulation. The result of this change to the design will delay the commencement of the project while each phase prior to phase C is revisited to include the missing requirements. The PM can reduce the risk of this issue occurring by following some of these best practices during schedule management:

- Include phase gate reviews in the schedule;
- Enforce the completion of these reviews prior to starting the next phase;
- Understand the level of maturity required of work products for each review;
- Document the entrance and exit criteria for all reviews;
- Ensure all work products for the milestone review is ready before entering the review;
- Document all feedback and track resolution within a change management system;
- Establish due dates for actions related to feedback;
- Progress to the next phase only after all project stakeholders have accepted the results of the milestone review and any corrective actions.

Milestone reviews are a valuable tool for the project manager and should be included in the schedule planning process. Allowing the schedule to take priority over meeting work product milestone review criteria will result in costly work stoppages and changes later in the project. For this reason, milestone reviews should not start until the work products are ready for the review, regardless of when the schedule calls for the review. The completion of products subject to review is always a predecessor to the start of a review and completed milestone reviews are always predecessors to the next phase of the project.

Level of Detail

Implementing standard top-level activities for work packages may not provide enough detail to ensure the work package is progressing as planned. At the same time, the actual performance of all the tasks to complete a portion of the work package can include thousands of very small items. One would not expect to track daily tasks such as, "locate file on network drive, open file in Microsoft Word, make necessary edits, or save file" as part of the project's master schedule. The PM strikes a balance between too little and too much detail.

The goal is to establish just enough detail in the schedule so that slippages in the schedule can be detected in time to recover from the slippage. There is a logical maximum and minimum duration for tasks depending on the scope of the work package and the duration to execute the entire work package. The following approaches can help the PM to strike the desired balance between too much and too little detail.

Ability to respond. The longest duration item within a work package should not exceed the amount of time the scheduled completion for the work package can slip. For example, a work package owner submits a plan with only three 1-month activities to complete a work package that has an overall duration of 3 months. The risk here is that by the time the first activity is reported late, the work package is already a month behind. It will be very difficult to make up a full month in the two remaining months of the work package. By breaking down the three activities for this work package into multiple tasks with durations of a week for each task, the maximum period of slip before detection is now 1 week.

Reporting time frame.: Many factors drive the PM's decision on frequency of reporting. These factors include overall duration of the project, level of risk exposure, political sensitivities, and executive-level interest in the project. However, requiring reporting intervals that are shorter than the major subtasks of a work package does not improve the PM's awareness. This level of detail may be of value to the work package owner but does not need to be tracked by the PM. The key here is not to require more detail than the reporting will support, leaving the lower level detail up to the work package owner.

Value of reporting. The level of effort to create, monitor, report, and maintain a schedule is directly proportional to the number of tasks in the schedule. Avoid spending more time on reporting status of tasks than the time required to complete the task.

Rolling Wave Plan

Earlier in this chapter the concept of progressive elaboration was presented as a means to start with a high-level project plan and then add increasing fidelity as the project team matures. This approach also applies to projects with extended durations or when an incremental development approach is followed. Just as with the initial schedule, some details beyond the near term may be left at a high level until the project progresses. Often, external factors outside of the control of the project may impact the ability to create an executable project plan for periods greater than a year. This case is particularly true in government-funded projects in which the budget is authorized on an annual basis. In the case of an incremental or spiral development approach, information learned in the near term is required before planning for the long term. In either example the schedule details can evolve as work is completed and the details of the future work becomes clearer. As with the initial schedule, care must be taken to avoid uncontrolled changes in project scope due to delayed schedule fidelity. In the event of rolling wave plans, the PM must ensure the key stakeholders (customer, PM's leadership, resource providers) are engaged in each periodic update to the schedule. The milestone reviews should always include a review of the updated schedule.

Invoicing Strategy

The completion of specific tasks and their deliverables is often used to trigger invoicing to the customer. For example, on a fixed price contract, the seller agrees to deliver products or services to the buyer for a fixed price. The value of the contract reflects the total price of delivering the entire scope of the project. To keep the project solvent, the agreement between buyer and seller allows for incremental invoicing based on the achievement of certain project milestones. It is important that the scheduled tasks tie back to the invoicing strategy so that funds flow into the project.

The PM uses the activity definition processes to establish a coherent means to communicate the project scope using a WBS and the project activity list. The WBS is the organizational structure for *what* is to be delivered. The activity list provides the details for *what* must be done to deliver the work described in the WBS. The PM uses work packages to strike a demarcation point of control between the individual resource pools owning each of the work packages and the PM. The WBS includes a mix of work directly associated with the end item product and indirect activities that support, manage, and lead the delivery of the direct efforts.

Sequence Activities

The sequence activities process involves defining the logical progression of the project. The primary goal of this step is to define the dependencies between activities and the order in which the activities are to be performed.

Activity Dependencies

Each task will have one or more predecessor and/or successor dependencies. Task predecessors are the one or more tasks that impact the ability to start or finish the given task. Task successors are those that can be impacted by the given task. With dependency phrasing, dependencies are always described from the point of view of the specific task. Figure 3.9 shows three tasks, each with a dependency on another task. The top statements reflect the relations between task B and the other two tasks. The bottom statements reflect the relationships for first task A then task C.

Dependencies between tasks can be tied to the start or completion of each task in the relationship. These dependency conditions specify the preconditions associated with the start or finish of the dependent tasks. The four most frequently used relationships are shown in Figure 3.10:

- Finish to start (task B cannot start until task A is complete);
- Start to start (task B cannot start until task A starts);
- Finish to finish (task B cannot finish until task A finishes);
- Start to finish (task B cannot finish until task A starts).

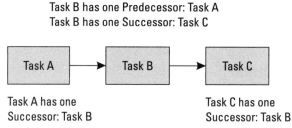

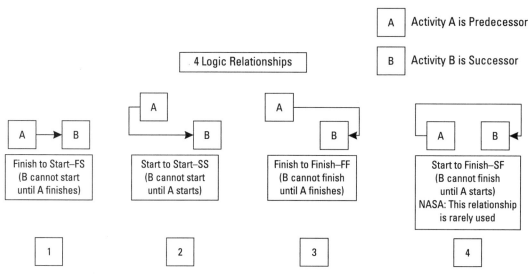

Figure 3.9 Task dependencies.

Figure 3.10 Common dependency types.

Each relationship can be further defined by adding a time frame for when a change in condition in the preceding task is applied to the succeeding task. A lag is added to a dependency to impose a wait period between the condition being met on the predecessor and when the successor responds to this condition. A lead allows the successive task to implement its response prior to the condition being met by the preceding task.

Network Diagrams

The sequences are defined by working from within each work package at its lowest level then rolling up to each parent task until the entire work package is sequenced. Dependencies with other work packages are defined and coordinated with other work package owners. There are several diagramming techniques available for creating a visual presentation of the sequencing and dependencies for the activities. Some of these techniques can be very detailed and complicated for new PMs to use. The key objective is to visualize the activities and their relationships. A simplified example is presented in Figure 3.11 to show the minimal data required to define basic dependencies.

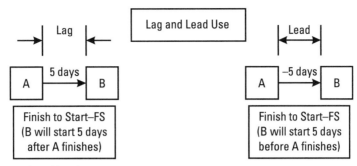

Figure 3.11   Lead and lag times.

Notice in Figure 3.12 sequencing and dependencies have been defined within the project management activities first that are then linked to external work packages or activities. The project starts off with the initiation and project management activities, which result in a project plan, charter, schedule, resources, and work authorizations. No work on the end item products can commence until a project management structure is established. In this example, the major subsystems are developed first with the subsystems coming together in the platform integration activity. The project then continues to the right with deployment, operations, and project closeout.

By including the subactivities of the project management activity in this diagram, the reader can see that the work authorizations task is why this activity is a predecessor to the rest of the project activities.

The sequence activities process is used to create the logical order of events within the project schedule. This logical order is driven by task dependencies in the form of preconditions for when a task may start, or finish based on the status of one or more other tasks. These dependencies are known as predecessors and successors to a given task. Predecessor and successive relationships can be further defined by adding time attributes that define when the dependent task responds to the condition in its predecessor task.

Estimate Activity Duration

With a clear definition of the activities and the relationships between these activities, the next step is to define the time required to complete each activity. During this process, each activity in the schedule is assigned an estimated duration to complete.

The planning described up to this point can be performed with a range of office or project management tools. The duration planning demonstrated herein can quickly grow unmanageable without a project scheduling tool. There are several desktop and web-based tools available that can store the working and nonworking days then calculate the elapsed time between the start and end dates based on these values.

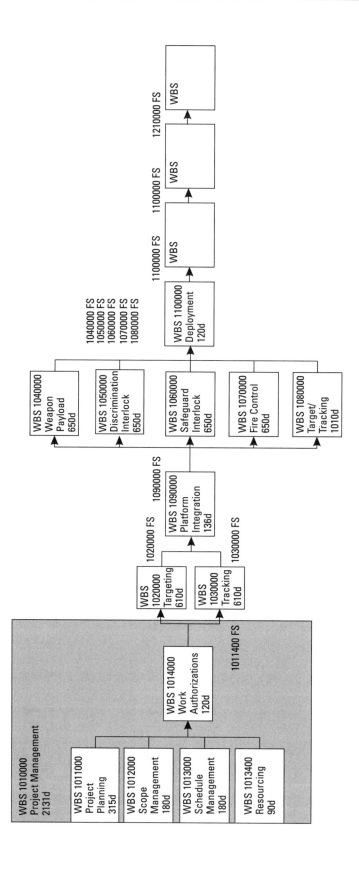

Figure 3.12 Example network diagram.

Duration and Elapsed Time

The activity duration is the actual amount of time spent working on an activity. All other time between the start and end dates is *wait time,* which is the sum of all nonworking periods including holidays, weekends, shutdowns, or nonproduction times.

The total elapsed time between the start and finish dates is the sum of the durations (working periods) plus any wait time (nonworking periods) (see Figure 3.13).

Elapsed time is not typically displayed in project tools which can cause some confusion when communicating the schedule. The PM must be clear on the difference between the duration of work and the total elapsed time when communicating activity durations. For example: the functional requirements task in Figure 3.14 has an elapsed time between the start and finish dates of 81 days and a duration of 60 days. When asked how long will the requirements task take? The answer is 81 days. When asked how much effort is involved? The answer is 60 working days for a single resource working on the task full time.

Start and Finish Dates

Start and finish dates in a project schedule are calculated based on: dependencies, durations, and hard constraints placed on the task. In a perfect scenario, the start

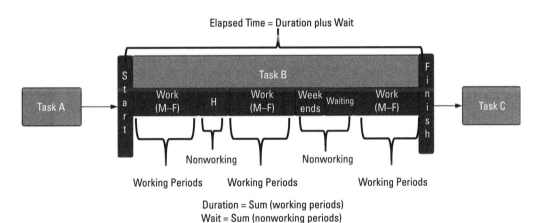

Figure 3.13   Total elapsed time for an activity.

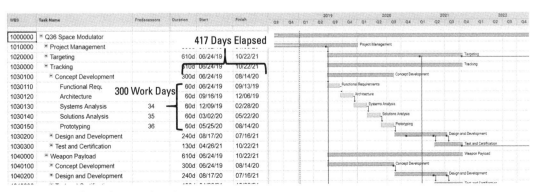

Figure 3.14   Elapsed versus duration.

or finish date is entered on a single activity in the schedule and all other dates are automatically calculated. This method is just a perfect scenario. Seasoned PMs recognize that creating perfect logic on a schedule with hundreds or thousands of lines can be time consuming and the goal is a realistic, achievable schedule not a perfect schedule. After all, schedules are estimates until the project is completed. Start and finish dates for each task are calculated based on

- *Predecessor dependencies*: Task B cannot start until task A is finished;
- *Successor dependencies:* Task B cannot finish before task C is started;
- *Start date constraints:* Task A must start no later than as specific date;
- *Finish date constraints:* Task A must Finish by a certain date and the duration drives the start date;
- *Duration plus wait time:* Based on a given start date, the finish date is after the duration and all wait times.

Activity scheduling constraints are explored in more detail in the "Schedule Analysis" section later in this chapter.

Activity Duration Rollup

During activity identification and activity sequencing, each element of the WBS was decomposed into smaller and smaller elements to define the activities that compose the project schedule. During estimate activity durations, the process works from the lowest levels of the WBS and rolls up each consecutive layer of the WBS, resulting in the project's overall duration (and elapsed time).

For example, in Figure 3.15 the concept development activity (WBS 103001) is comprised of five sequential tasks each with a duration of 60 days each. These individual durations roll up to a duration of 300 days for the parent activity.

Activity duration rollup can also account for any overlapping or concurrent tasks. In Figure 3.15 the design (WBS 103020) is comprised of six overlapping

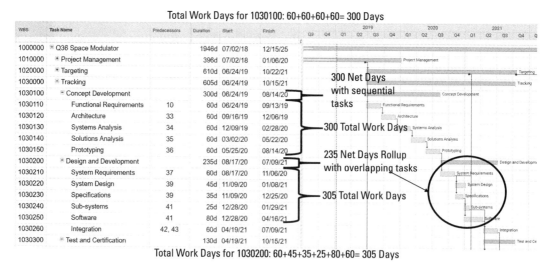

Total Work Days for 1030100: 60+60+60+60= 300 Days

WBS	Task Name	Predecessors	Duration	Start	Finish
1000000	Q36 Space Modulator		1946d	07/02/18	12/15/25
1010000	Project Management		396d	07/02/18	01/06/20
1020000	Targeting		610d	06/24/19	10/22/21
1030000	Tracking		605d	06/24/19	10/15/21
1030100	Concept Development		300d	06/24/19	08/14/20
1030110	Functional Requirements	10	60d	06/24/19	09/13/19
1030120	Architecture	33	60d	09/16/19	12/06/19
1030130	Systems Analysis	34	60d	12/09/19	02/28/20
1030140	Solutions Analysis	35	60d	03/02/20	05/22/20
1030150	Prototyping	36	60d	05/25/20	08/14/20
1030200	Design and Development		235d	08/17/20	07/09/21
1030210	System Requirements	37	60d	08/17/20	11/06/20
1030220	System Design	39	45d	11/09/20	01/08/21
1030230	Specifications	39	35d	11/09/20	12/25/20
1030240	Sub-systems	41	25d	12/28/20	01/29/21
1030250	Software	41	80d	12/28/20	04/16/21
1030260	Integration	42, 43	60d	04/19/21	07/09/21
1030300	Test and Certification		130d	04/19/21	10/15/21

Total Work Days for 1030200: 60+45+35+25+80+60= 305 Days

Figure 3.15 Task duration rollup.

tasks with durations of 60, 45, 35, 25, 80, and 60 days, which totals up to 305 days. However, in this case the scheduling software accounts for the overlapping task and calculates a total duration for the design task as 235 days.

Resource Loading

At its basic definition, days in a duration means one day of resource with a set number of hours (typically 8 in the United States). The number of people working on the task does not impact the number of days required. This level of planning is acceptable for the average project.

However, PM tools can calculate the value of a single day of duration based on the number of shifts worked per day, the length of shifts, and the number of resources assigned to each task shift. In these cases, the duration and its unit of measure (days/weeks/hours) is defined as a specific amount of resources working a specific number of hours per day. This level of detail is often required when developing production estimates or labor costs for high-volume work. In these scenarios, the duration is calculated versus manually entered. This level of planning logic can be very complicated and is recommended for only when the resource loading is used to calculate or track the cost of work, such as in earned value management (EVM). For more information on EVM and cost estimation and tracking, see Chapter 8 within this book.

Duration Estimating Techniques

Three common techniques for estimating the duration of project activities are: expert judgment, analogous estimating, and three-point estimates. The PM is not expected to be an expert in every aspect of the project. A PM relies on the subject matter expertise throughout the schedule management process and estimating duration requires subject matter expertise to support all the methods provided herein. Two or more of these methods can be used in conjunction to increase the level of surety in the estimates.

Expert judgment (subject matter expert): Expert judgment is one of the least expensive and fastest methods of estimating task duration. A person with expertise in the field of work associated with the task provides an estimate for its duration. As this method does not require the analysis of large amounts data or interaction with multiple parties, the method is both quick and inexpensive.

The largest risk associated with using this method is relying on a person who lacks the required qualifications. A qualified person is someone with relevant and recent experience in performing the associated work on multiple recent projects. Senior managers that have been away from the direct work for extended periods may have limited understanding of the current techniques. This case is especially true in technology and scientific fields where the state of the art changes rapidly.

Unfortunately, the best talent is not always available when needed and this is particularly true during the initiation phase. This talent is often reserved for the real work. Caution should be used with the less experienced estimator as overoptimism is common. The same holds true with the overly pessimistic views of the old timers who have seen too much and may be a little less adventurous at this point in their careers. Using two or more experts in this process can help find the median between these two extreme views.

Analogous estimating (historical data–actuals): Analogous estimation uses data from previous projects to create estimates of future work. The estimates created with this method benefit from being based on actual data instead of being a purely theoretical exercise or one's opinion. However, one will be surprised to learn that few projects are identical in complexity, constraints, and influences, which means the level of similarities may be limited. This reality does not mean that analogous methods are ineffective. What is does mean is that care must be taken to understand and document the similarities and differences between the historical data and the current project. With this understanding, other methods can be used to validate the results of the analogous estimate to help reduce the level of uncertainty. When this method is augmented with expert judgment, additional accuracy can be obtained while retaining its foundation in previous project data. There are inherent issues with historical data that must be considered:

- Past project data is often corrupted by undocumented hours and complications in the prior work.

- Salaried employees are often restricted from documenting actual hours in the project billing system.

- Project management data is often disconnected from the organization's financial system data. Therefore, the project management data may only reflect the level of effort planned for an activity and may not include the number of hours worked on specific tasks.

- In times of urgency, work hours are extended to maintain schedule. These hours are not always documented in the project management database.

- Few human resources are the same. Two employees with the similar titles and seniority can have two different levels of experience and skills.

Weighted three point estimating. The program evaluation review technique (PERT) is a method that attempts to account for the uncertainty in estimating future tasks when historical data or expert opinion is limited. This method can help to assess the quality of an estimate provided by a less experienced team member or an overly pessimistic old timer.

The estimate starts by identifying the best-case, worst-case, and most-likely values for the activity duration. A good rule of thumb for most-likely is that the task has been done numerous times with similar results. A worst-case scenario is based on the occurrence of the highest impact risks identified for this task and best-case is based on the occurrence on the highest impact opportunities identified.

A threat is an unplanned event that may negatively impact the task. An opportunity is a form of risk that can positively impact the task.

The PM works with the estimator to develop an estimate based on what they deem to be the most likely duration. Next the PM asks the estimator to identify how long the task would take if the identified risks occurred (worst case). Lastly the PM asks the estimator to identify the duration if everything went well or the opportunities identified were to occur (best case).

Using these three numbers, a three point estimated value can be calculated using basic averages or weighted average calculations as is common with PERT analysis.

The equation for a basic average is

$$\text{Estimated duration} = \frac{\text{best } + \text{most likely} + \text{worst}}{3}$$

In basic averaging, the probability of any of the three scenarios occurring is treated as equal.

A weighted average equation is used when the probability of one or more events occurring does not match that of the others.

The equation for a weighted average is

$$\text{Estimated duration} = \frac{\text{best}}{6} + \frac{4 * \text{most likely}}{6} + \frac{\text{worst}}{6}$$

This weighting (the 4) is drawn from a triangular probability distribution that gives more weight to one value in the calculation as is common in PERT analysis. The equation puts more weight on the most likely case and equal weighting on the best and worst cases. In other words, the probability of the most likely is greater than the other two estimates and the other two estimates are equally less probable than the most likely of occurring. In a hypothetical scenario the electrical engineer provides the following most likely = 7, best = 5, and worst = 14.

With unweighted averages the following is revealed:

$$\text{Estimated duration} = \frac{5 + 7 + 14}{3} = 8.67$$

Weighted average calculation using the same estimates results in:

$$\text{Estimated duration} = \frac{5}{6} + \frac{4 * 7}{6} + \frac{14}{6}$$

$$\text{Estimated duration} = .833 + 4.67 + 2.33 = 7.83$$

In this example, the result of the weighted distribution is a lower estimated duration. This difference is because the most-likely value is increased by 4 times, shifting the new estimate closer to the most-likely scenario.

Scheduling does not always follow a normal probability distribution between the three values. Therefore, other distributions can be applied in these cases.

However, that topic is best left for an advanced study on probability distributions for project estimation. The point to take away from this discussion is that three-point estimating is a good technique to reduce uncertainty due to limited historical data and can be used in combination with the other methods described herein to increase the level of confidence in activity duration estimates.

Reserve Analysis

Reserve analysis is used in conjunction with the other techniques to identify the amount of buffer needed across the project for use when risks are realized. Reserve can be applied across the totality of the project or at the work package level. In the first case, the leadership of the project organization elects to reduce the due date for the project by a certain percentage of the overall project schedule. In the other application, the leadership authorizes the delegation of a portion of the reserve to elements of the schedule that are deemed to have the highest risk exposure to the schedule. This reserve time in the schedule is meant to account for all known risks during the project.

The PM relies on the subject matter expertise to create estimated durations for each task. These estimates are calculated using one or more of three methods: expert judgment, analogous estimating, and three-point estimates. Reserve analysis is used in conjunction with the other estimating techniques to identify the amount of slack or schedule reserve required for the project. This reserve can be held back until required at the time or risk occurrence or allocated directly to the highest risk activities.

Estimate Activity Resources

At this point in the scheduling process, the PM has a clear set of activities, defined relationships between these activities, and preliminary durations defined for each activity. The *what* and *when* have been defined. Now the PM determines *who* is required to execute the work. The goal of the process is to define the resources required to complete the project activities while ensuring efficient and effective use of these resources. Resources include both the human resources (people, contractors, vendors) as well as the physical resources (equipment, tools, software) as shown in Figure 3.16.

Resource requirements define the talents and capabilities needed for each task, the quantity of these talents, and when these talents are needed. The need for subject matter experts remains in this stage as it did in all prior stages.

It is very important to recognize that the typical project manager has very little authority over their resources. In many cases, the coworkers assigned to the project

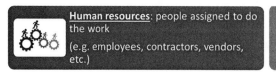

Figure 3.16   Resource types.

have no supervisory relationship with the PM as they report to a different line manager in the organization. Also, very few organizations allow the PM to make contractual agreements for the company. This authority to establish and change contracts is reserved to a procurement officer in the organization. Therefore, in either case, the onus is on the PM to establish and communicate resource requirements and performance expectations to build a successful team.

Resource Requirements

The method for defining the resource requirements can follow two major approaches:

- Work package delegation, where packages of work are delegated to individual departments, contractors or managers. In this method, the work package owner provides an estimate of resources required to support each work package. The work package owner uses his/her resources across multiple tasks and can surge in and out of the project as needed.
- Individual Recruitment, where the PM must clearly define the position description for each group of tasks and then recruit or negotiate the assignment of individuals joining the team.

There is no perfect answer or best strategy for identifying and recruiting resources. In fact, each project may have a combination of both approaches for elements of the project. The PM improves the likelihood of making sound decisions by understanding the project scope, identifying the unique characteristics of each work package and communicating this information to his/her resource providers. The key here is to establish a clear understanding of the types and quantities of resources required. How resources are recruited and obtained is further explained in Chapter 6.

Develop Schedule: Analyze and Baseline

As mentioned at the opening of this chapter, the schedule is an amalgamation of the data points collected by following the steps described thus far. With the activity list as the foundation, the subsequent processes add the data necessary to create a schedule. But the schedule creation process is not yet complete. The schedule must be reviewed for completeness, errors, logical flow, feasibility, and consistency with the project constraints.

Schedule Analysis

Working with the project stakeholders, subject matter experts, and resource providers the PM and scheduler review the following data points in the schedule:

Alignment to major milestones. The project schedule is checked to verify that the major milestones identified within the contract or during activity identification are met within the expected timeline.

Schedule constraints. The project schedule is verified that all constraints placed on the project by the customer and the project manager's organization are satisfied within the expected timeline. Schedule constraints can be applied to individual tasks and can override any restrictions placed on the start and finish dates by predecessors. There are eight types of constraints:

- As late as possible;
- As soon as possible;
- Finish no earlier than;
- Finish no later than;
- Must finish on;
- Must start on;
- Start no earlier than;
- Start no later than.

Using constraints is not wrong, but they are an indicator of hard deadlines or inflexibility in the schedule.

Calendar constraints. Check the calendar to ensure none of the major activities, deadlines, or events coincide with national holidays or work stoppages. It is important to look to see if a major milestone review occurs during a period when people are known to take time off, employees are scheduled for training, or a production facility is closed.

Working hours. When using resource loading, it is important to verify that the schedule's calendar is set up to account for any special shift work (first, second, and/or third shifts) or rotating shift schedules.

Scope management. The schedule as composed should be reviewed to ensure it did not modify the project scope. This step involves comparing all activities with their parent WBS element to ensure that working being performed under a work package is consistent with the scope of the work package. Additional fidelity to the work is good; additional scope to the work is change.

Verify sequences. Using network analysis, the data is checked to identify any activities with no linkage to the rest of the project. This condition means one of two things: the activity is not properly linked to its dependents or the activity is not necessary for the completion of the project (scope creep).

Resource leveling and availability. Resource assignments are checked to ensure that resources are not tasked beyond their availability. Availability is based on level of commitment (full-time/part-time), planned outages (training, shop closures, vacations), and commitments to other projects.

Reserve allocation/over allocation. The allocation of reserves is verified to ensure any reserve assigned within the schedule has been authorized by leadership. It is also wise to check for patterns of overly pessimistic schedules where slack has been added at multiple levels of the work package's schedule.

Resource planning. The schedule is reviewed with all resources providers to achieve concurrence and commitment to support the project schedule. This is a good time to verify that hiring or infrastructure investments are underway and have adequate leadership attention and support.

Customer and management concurrence. The schedule is a critical data point for the project and the PM must ensure that both the customer and the PM's leadership have reviewed, understand, and concur with the schedule.

At the opening of this discussion on schedule management, the idea was presented that the project schedule is the result of negotiations between the competing interests of the customer, the PM's leadership, resource sponsors, and available resources. It is good to assume that some rework and trade-offs will be required to achieve a schedule that depicts an acceptable plan to execute the scope. The schedule is a critical tool for leading the project team and is a living data set. As such, schedule development is not complete until the project is complete.

Schedule Baselines

As stated before, the schedule is a critical data set for the project. In as such, the schedule must be managed under strict configuration control. Upon concurrence from the project owner, resource providers and the project team, the schedule is baselined and becomes a configuration item for the project (see Chapter 9).

A baseline is a snapshot of the project schedule that establishes the planned schedule for the project as shown in Figure 3.17. In practice, the data within the duration, start date, and finish date fields is copied to a new set of columns and changes are no longer made to these values.

The schedule will change, this fact is inevitable. But any changes to the planned schedule are managed through a deliberate and controlled process, resulting in the creation of a new baseline. A new planned schedule baseline does not overwrite any previously saved baselines. Each new baseline is saved separately as a record of the project's history. This method will build important historical data for the current project while providing analogous data for future projects.

The schedule is an amalgamation of the data points collected by following steps to identify the activities (the what), determining their duration and sequencing (the

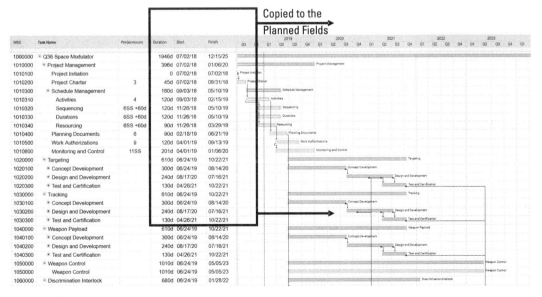

Figure 3.17   Baselining the schedule.

when), and identifying the resource qualifications for these activities (the who). With the activity list as the foundation, the subsequent processes add the data necessary to create the schedule. The schedule creation process verifies the data in the schedule matches the scope of the project and validates the schedule is reasonable. With concurrence from the project stakeholders, the schedule is baselined as a configuration management item.

Use the Schedule

By following the steps leading up to this point, the PM now has a reasonable plan to execute the project. The plan is an estimate for the activities, durations, and resources required. Novice PMs may be tempted to celebrate this milestone and put all this scheduling work behind them. Any time that a PM treats the schedule as a project deliverable that once done can be put on a shelf, he or she wastes a lot of time and money and the project will likely fail. The schedule development and management processes continue until the project is closed out. The schedule is the project's roadmap (the plan); it is used to communicate the plan, to monitor progress according to the plan, to identify potential hazards along the way, and to define potential detours.

Tracking Progress

During project execution, each attribute of the schedule (start, finish, duration) has two or more values. The planned values are the estimated values created during the planning process and copied to a baseline. A baseline is a snapshot of the project schedule that establishes the planned schedule for the project. The baseline fields do not change from this point forward. As tasks are started and finished the actual dates for these events are entered in the original estimated fields for each activity. Actual start and finish *v*alues that differ from the planned values will impact the estimated start and finish dates downstream based on their dependencies.

This concept can be confusing. A good way to visualize the schedule data life cycle is to explore the way a Global Positioning Service (GPS) operates during a trip.

Prior to starting a trip, a driver enters the starting and ending points into the system. The system then calculates the recommended path or paths based on known data. Upon selection of the desired path, the driver instructs the GPS to begin guiding the journey. The GPS then calculates the estimated distance (duration) and the expected travel time (elapsed time), and the projected arrival time (project finish date). The path is saved into the GPS memory (baselined) and is used as the planned route from that point forward [2].

As the vehicle makes progress, the GPS informs the driver each time a new action is required, such as turning, exiting, or navigating an interchange (initiation of new work packages). The GPS continually tracks the progress made (actual values) and uses this data to create new estimated durations, elapsed time, and expected arrival time. If the driver makes faster than planned progress, these values are reduced and if the progress is slower than planned, the values are increased (revised estimated values) [2].

Like the GPS, the PM monitors progress by using reportable status attributes in addition to the actual start and finish dates. These attributes provide indicators on the progress of an activity between its start and finish dates. There are many ways to collect and report activity performance. The two most common are tracking the percentage of work completed and the status of the activity (not started, started, completed). These two attributes provide the PM a quick indication of current activity and progress [2] (see Figure 3.18).

Assessing Performance

Status indicators provide insight into individual tasks, but they do not provide indications of performance patterns nor do they alert the PM of potential downstream effects. As with status monitoring, there are a multitude of techniques for measuring and indicating performance. At a minimum the PM requires the ability to detect downstream impacts and the ability to measure the difference between the planned schedule and the actual schedule. This later measurement is referred to as variance, which is the delta between a planned value and the actual value. Variance can be positive (faster than planned) or negative (slower than planned). This functionality can be created in a spreadsheet tool, but project management tools are better equipped. Figure 3.19 shows the project schedule with the addition of two new columns: start variance and finish variance.

Also shown in Figure 3 19 are the dates that are impacted (highlighted) when an activity finishes later than planned. In this case the initial activity definition task for the schedule (WBS 1010310) was completed 14 days late. The late completion of this one event has an impact on many downstream events that were automatically highlighted by the scheduling tool. Without the scheduling tool's conditional formatting, the impact of this delay early in the project may go unnoticed.

Analyzing Performance

The performance indicated in the schedule's current progress should raise notice, but it may not need to raise panic. In the event of a slippage on one or more tasks,

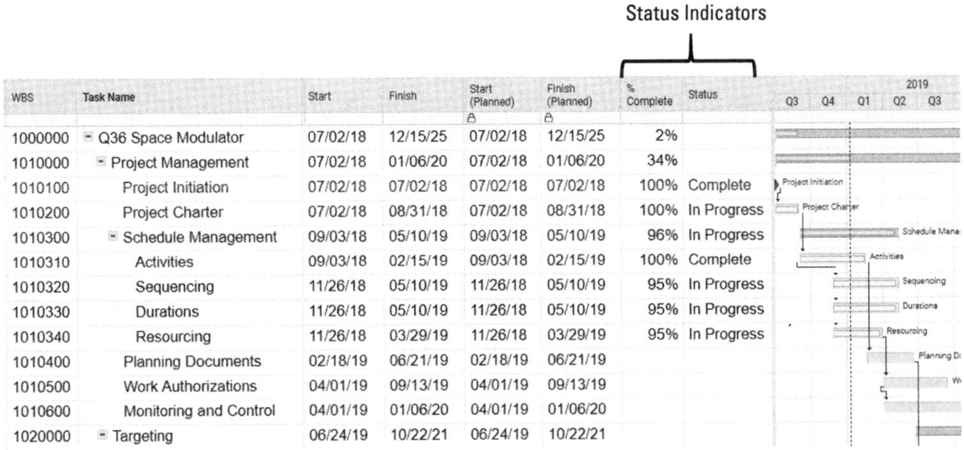

WBS	Task Name	Start	Finish	Start (Planned)	Finish (Planned)	% Complete	Status	Q3	Q4	Q1	2019 Q2	Q3
1000000	Q36 Space Modulator	07/02/18	12/15/25	07/02/18	12/15/25	2%						
1010000	Project Management	07/02/18	01/06/20	07/02/18	01/06/20	34%						
1010100	Project Initiation	07/02/18	07/02/18	07/02/18	07/02/18	100%	Complete					
1010200	Project Charter	07/02/18	08/31/18	07/02/18	08/31/18	100%	In Progress					
1010300	Schedule Management	09/03/18	05/10/19	09/03/18	05/10/19	96%	In Progress					
1010310	Activities	09/03/18	02/15/19	09/03/18	02/15/19	100%	Complete					
1010320	Sequencing	11/26/18	05/10/19	11/26/18	05/10/19	95%	In Progress					
1010330	Durations	11/26/18	05/10/19	11/26/18	05/10/19	95%	In Progress					
1010340	Resourcing	11/26/18	03/29/19	11/26/18	03/29/19	95%	In Progress					
1010400	Planning Documents	02/18/19	06/21/19	02/18/19	06/21/19							
1010500	Work Authorizations	04/01/19	09/13/19	04/01/19	09/13/19							
1010600	Monitoring and Control	04/01/19	01/06/20	04/01/19	01/06/20							
1020000	Targeting	06/24/19	10/22/21	06/24/19	10/22/21							

Figure 3.18   Status indicators.

One late activity impacts multiple activities

WBS	Task Name	Predecessors	Duration	Start	Finish	Duration (Planned) A	Start (Planned) A	Finish (Planned) A	% Complete	Status	Start Variance	Finish Variance
1000000	⊟ Q36 Space Modulator		1956d	07/02/18	12/29/25	1946d	07/02/18	12/15/25	2%		0	-14
1010000	⊟ Project Management		396d	07/02/18	01/06/20	396d	07/02/18	01/06/20	34%		0	0
1010100	Project Initiation		0	07/02/18	07/02/18	0	07/02/18	07/02/18	100%	Complete	0	0
1010200	Project Charter	3	45d	07/02/18	08/31/18	45d	07/02/18	08/31/18	100%	In Progress	0	0
1010300	⊟ Schedule Management		180d	09/03/18	05/10/19	180d	09/03/18	05/10/19	96%	In Progress	0	0
1010310	Activities	4	130d	09/03/18	03/01/19	120d	09/03/18	02/15/19	100%	Complete	0	-14
1010320	Sequencing	6SS +60d	120d	11/26/18	05/10/19	120d	11/26/18	05/10/19	95%	In Progress	0	0
1010330	Durations	6SS +60d	120d	11/26/18	05/10/19	120d	11/26/18	05/10/19	95%	In Progress	0	0
1010340	Resourcing	6SS +60d	90d	11/26/18	03/29/19	90d	11/26/18	03/29/19	95%	In Progress	0	0
1010400	Planning Documents	6	90d	03/04/19	07/05/19	90d	02/18/19	06/21/19			-14	-14
1010500	Work Authorizations	9	120d	04/01/19	09/13/19	120d	04/01/19	09/13/19			0	0
1010600	Monitoring and Control	11SS	201d	04/01/19	01/06/20	201d				The schedule	0	0
1020000	⊟ Targeting		610d	07/08/19	11/05/21	610d				variance indicates	-14	-14
1020100	⊟ Concept Development		300d	07/08/19	08/28/20	300d				activities that may	-14	-14
1020110	Functional Requirements	10	60d	07/08/19	09/27/19	60d				start and/or finish	-14	-14
1020120	Architecture	15	60d	09/30/19	12/20/19	60d				14 days later than	-14	-14
1020130	Systems Analysis	16	60d	12/23/19	03/13/20	60d				planned	-14	-14
1020140	Solutions Analysis	17	60d	03/16/20	06/05/20	60d					-14	-14
1020150	Prototyping	18	60d	06/08/20	08/28/20	60d					-14	-14
1020200	⊟ Design and Development		240d	08/31/20	07/30/21	240d	08/17/20	07/16/21			-14	-14
1020210	System Requirements	19	60d	08/31/20	11/20/20	60d	08/17/20	11/06/20			-14	-14
1020220	System Design	21	60d	11/23/20	02/12/21	60d	11/09/20	01/29/21			-14	-14
1020230	Specifications	21	60d	11/23/20	02/12/21	60d	11/09/20	01/29/21			-14	-14
1020240	Sub-systems	23	60d	02/15/21	05/07/21	60d	02/01/21	04/23/21			-14	-14
1020250	Software	23	60d	02/15/21	05/07/21	60d	02/01/21	04/23/21			-14	-14
1020260	Integration	24, 25	60d	05/10/21	07/30/21	60d	04/26/21	07/16/21			-14	-14

Figure 3.19   Performance indicators.

the PM determines the impact level on the entire schedule. In this example, the PM can quickly see that the projected end date for the overall project has slipped by 14 days. Does this mean the project is doomed or that the major milestones will not be met? Further analysis is required to make these determinations.

Critical Path Analysis

The critical path is the longest path in the schedule to complete the project. The critical path represents the minimum duration of the project. For example, Figure 3.19 demonstrates how a delay in the completions of one activity (WBS 101030) has impacted the end date of the project by 14 days. Therefore, this task is in the critical path. However, the PM does not yet know if the project is as bad as it looks in Figure 3.19. The PM does not know if the recovery plan must address all downstream tasks (highlighted in Figure 3.19) or only a few of the tasks. Luckily the PM has selected a scheduling tool that can automatically calculate and show the critical path. Figure 3.20 shows the critical path calculated by the tool in which the Gantt bars of the affected activities are highlighted and connected by a dark dashed line. Notice that not all the Gantt bars for delayed tasks are not highlighted. This lack of Gantt bar highlighting indicates that these tasks, even though impacted, can support some delay before impacting the project's critical path. This information gives the PM insight to the activities that may need to adjust due to this delay.

Further analysis of the schedule by following the critical path lines down the schedule indicates that there is a critical dependency between the system engineering activities inside of two different subsystems. In Figure 3.21 the design of the targeting system must be completed before the design of targeting/tracking interface for the fire control system can proceed.

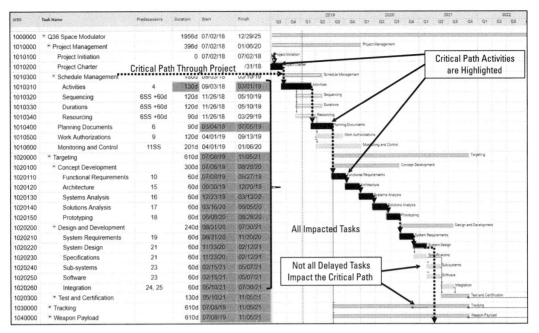

Figure 3.20 Critical path analysis.

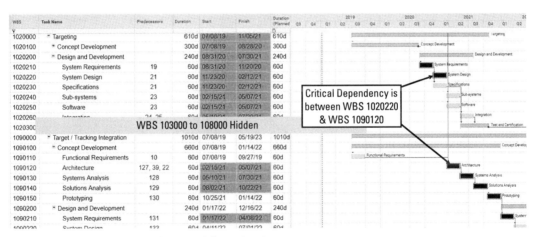

Figure 3.21 Identification of the critical dependencies.

The delay in one activity early in the project has impacted 60% of the activities in the schedule. Critical path analysis has narrowed the search down to two critically dependent tasks. Now the PM can focus on the steps necessary to recover by focusing first on the impacted downstream events.

Corrective Actions

The PM has several courses of action to follow when responding to issues with the project plan. The factors that influence the PM's decisions here are the same as those that influenced the initial schedule development: The balance between cost, scope, and schedule. Each of the following approaches can result in a shorter schedule. The challenge is determining the acceptable balance of the three drivers.

- *Schedule compression.* Schedule compression involves shortening activity durations within the critical path without changing the scope. The most common two ways of shortening the schedule are: crashing or fast-tracking.

- *Crashing.* With schedule crashing, the durations of one or more activities in the critical path are shortened by increasing the amount of resources applied to the activities. The resource increase can be achieved by adding resources or by increasing the workday of the existing resources. In either case, the cost to apply additional resources will impact the overall cost of the activity. In the current example this response may involve adding resources to one or both critically dependent activities (tracking system design, and targeting and tracking design).

- *Fast tracking.* Fast tracking changes the sequencing of sequential activities, so they overlap or run in parallel. Fast tracking can include starting a dependent task prior to the completion of its predecessor. For example, in the case of the dependency between the two system designs, it may be feasible to start the target and tracking Integration design (WBS 1090120) a few weeks earlier based on the preliminary results of the tracking system design (WBS 1020220). This approach should not directly increase cost but there will be a new risk for rework based starting the target and tracking integration design prior to the tracking system design being completed.

- *Reduce scope.* Reducing scope to overcome delays is rarely the preferred option. However, technological challenges arrive in system development projects that may be insurmountable or infeasible to resolve. The urgency to improve existing capabilities may mean a less-than-perfect solution can be accepted. For example, there was a program within the U.S. military to build a vehicle that could resist the impact of mines and improvised explosive devices (IEDs). Military forces where being maimed and killed every day in the battlefield. A performance parameter for braking at high speeds could not be met. However, the important capability of protecting forces while in transit was being satisfied. In this case, the military elected to accept a safe but slower vehicle until the braking speeds could be improved. In the current example, removing the design efforts from the project is probably not a good solution.

- *Accept the delay.* Accepting the delay may result in a delay in the schedule. If the delay is accepted by all stakeholders, a new baseline (as planned v2) is saved from that point forward. Accepting delay and extending the total duration of the schedule will cost more money. Costs are incurred every day the project office and its resources are assigned to the project.

- *Wait and see.* Wait and see does not mean ignore the problem and hope it goes away. Wait and see acknowledges the current delay is a very small percentage of the overall 600-day schedule and chances are other events will unfold between the current date in time and when the delayed critical path task is planned to start.

The best approach to addressing project performance varies from project to project and even from activity to activity. An approach that worked on one case may not be well suited for a very similar case. The key is to understand where the problem lies in the schedule and focus efforts on the area of critical dependency. The point of critical dependency is not always the point where the schedule slipped. It could be much further downstream from the source. By analyzing each option using the cost, schedule, and scope triangle the PM can make informed recommendations to the project stakeholders regarding corrective actions.

Summary

This chapter provided a six-step process for the development, approval and management of the project schedule.

The schedule is an amalgamation of the data points collected by following steps to identify the activities (the what), determining their duration and sequencing (the when) and assigning qualified resources to these activities (the who).

The project schedule is a critical tool for successful project management and must be understood and utilized by the entire project team. The project schedule is an estimate based on the project team's current understanding of the project and its constraints. The project schedule starts off very high-level and is progressively elaborated as the project team's understanding improves. Therefore the project planning and management process endures the entire project life cycle.

References

[1]   National Aeronautics and Space Administration (NASA), *NASA Schedule Management Handbook*, Washington, DC: NASA, 2010.

[2]   Lillard, J. D., "Capability Assurance Instititute Blog," September 20, 2018 [online], http://capabilityassurance.com/blog/.

[3]   Project Management Institute, Inc., *A Guide to the Project Management Body of Knowledge (PMBOK Guide)*,Sixth Edition, Newtown Square, PA: Project Management Institute, Inc., 2017.

[4]   National Aeronautics and Space Administration, *NASA Systems Engineering Guide*, Washington DC: National Aeronautics and Space Administration, 2007.

CHAPTER 4

Requirements Management

At the most basic level, a requirement is something that is needed or is compulsory. In systems development, a requirement's maturity or validity is measured with characteristics such as completeness, singularity, clarity, necessity, and uniqueness. Below are several industry publications that define requirements engineering and management. Requirement management involves understanding what product or service to deliver to satisfy the contract. To accomplish this, requirements management involves identifying all the customer "shall" statements, decompose requirements into lower-level specifications, specify verification method, and validate them through requirements verification traceability matrix to deliver the product or service per customer and contract expectations.

> *The International Organization for Standards (ISO) 15288:2015 System and software engineering – Systems lifecycle processes establishes a common lexicon for describing the activities associated with requirements for systems development [1].*
>
> *The International Council of Systems Engineering (INCOSE): Systems Engineering Handbook provides guidance on the identification, evolution and management of system requirements as defined within 15288:2015 [2].*
>
> *The International Council on Systems Engineering (INCOSE): Guide for Writing Requirements (2017) provides specific guidance on how to express requirements in the context of a Systems Engineering endeavor [3].*

The goal of this chapter is to expose the PM to the concepts of requirements management as an important activity required for successful project management. Requirements management is important to the project manager because it establishes and maintains a common understanding of the product or service that must be delivered to satisfy the project's charter.

> *Project requirements start with what the user really needs (not what the provider perceives that the user needs) and end when those needs are satisfied.*
> —*Visualizing Project Management* by Forsberg, Moog, Cotterman [4].

As shown in Figure 4.1, the requirements management process starts with the identification of a need (stakeholder, organization, customer, or user). These needs are then transformed into a set of requirements through a recursive process of collection, elaboration, and analysis. Requirements are expressed in the form of structured statements that communicate these needs and expectations to the parties

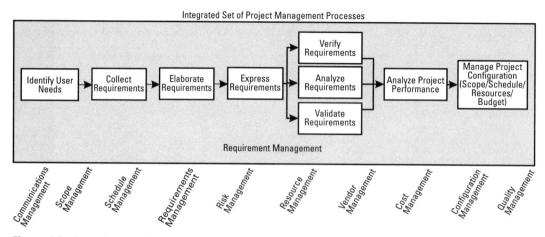

Figure 4.1   Recursive requirements management process [3].

charged with developing a solution. The resulting requirements are verified to be written according to accepted practices and then validated as being consistent with the customer needs and expectations. Throughout this process the requirements are traced from their higher-level need statements and source documents down through the system design.

Requirements are the single thread that goes through a project life cycle from its conception through design, build, test, and customer acceptance. A lack of clarity in the project requirements equates to a lack of clarity in the project scope. As with changes in scope, the cost to respond to a change in requirements increases as the project matures.

According to PMI's Pulse of the Profession (2014) study, 37% of all organizations surveyed reported inaccurate requirements as a leading factor for projects that failed. PMI also found that 87% of all project organizations that participated in the study acknowledged the need to improve their application of project management [5].

Requirements changes are inevitable in a project and establishing robust requirements change management structure helps to ensure the impact of each change is understood and approved by the project stakeholders. Every change in requirements is a change in project scope. As shown in Figure 4.2, each requirement has an associated impact on the cost, schedule, and performance of the project.

The customer may not be the only potential source of change. The PM may find the team being overly excited to improve product capability. However, the project team must understand that any change will impact the project. The PM and project technical lead must ensure the delivery of only what is specified in the project scope and requirements or face cost and schedule overruns.

Document what's agreed upon, deliver what's agreed upon, and do no more.
—Eric J. Roulo

Project team. Table 4.1 defines the roles and responsibilities associated with the processes described within this chapter.

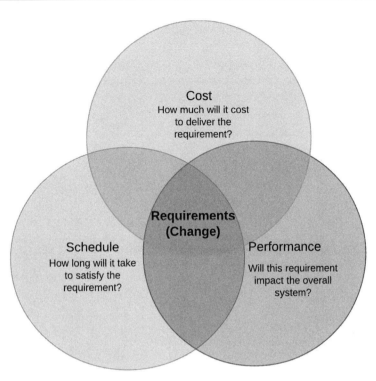

Figure 4.2 Impact of change on cost, schedule, and performance.

Table 4.1 Roles within Requirements Management

Role	Responsibility
Customer lead	The consolidated voice for requirements from the customer's organization. Has final authority over all additions, changes, or deletions to requirements.
Project manager	The person accountable for project success, the PM leads, monitors, and supports the processes and procedures associated with requirements identification and management.
Customer technical representatives	Charged with ensuring the customer's standards for safety, security, interoperability, and strategy are included in the requirements and design.
Customer user representatives	Members of the customer's organization that perform the business or mission functions to be supported by the project's outcome (system, service, process).
Engineers/analyst	Members of the project manager's technical team that collect, elaborate, express, and analyze requirements. All approved requirements are then satisfied within the design.
Requirements manager	Member of the project manager's team that performs the consolidation, structuring, publication, and management of requirements. May be performed by the engineers and analysts as a secondary function or by a dedicated resource.
Procurement officer	Member of the project manager's team that can authorize changes to the agreements between the project team and the customer or between the project team and its vendors.

The project manager works with the team and customer to define clear and concise requirements, manage change control, collaborate changes with all key stakeholders, and maintain solid traceability of requirements throughout the project life cycle from design to, build, testing, and delivery. The PM maintains an agreement with the customer and project technical lead on the requirements every step of the way until the product is accepted in accordance with these requirements.

During requirements management, requirements are refined and matured through the iterative process depicted down the left side of Figure 4.3 and explained in the following sections. The information identified in each step of this process constitutes requirements. What changes is the level of detail and specificity described in the requirements.

Publications vary in the terms used to describe requirements at each layer abstraction. For this discussion the terms and definitions used herein are provided in Table 4.2.

The following sections of this chapter provide an overview for the activities and techniques associated with each of the steps shown in Figure 4.3. The project to deliver the Q36 Illudium Space Modulator for Marvin the Martian is used to create example requirements [6].

Identify User Needs

The objective of this step is to document the problem or capability gap to be solved by the project. Sound requirements can be traced from the earliest source document through the design process to the final deliverable. The project's initiating documentation includes background information on the project, a definition of the customer need, a description of the desired solution, and the conditions by which the solution is to be designed, developed, and delivered.

Projects exist to satisfy a customer need. These needs are described in the project charter, project objectives, and statement of work. During scope management, the project's scope was described in the form of deliverables within the WBS. Requirements serve to describe each deliverable, product, service, or outcome to enough degree they can be specified, designed, constructed, delivered, and accepted. As

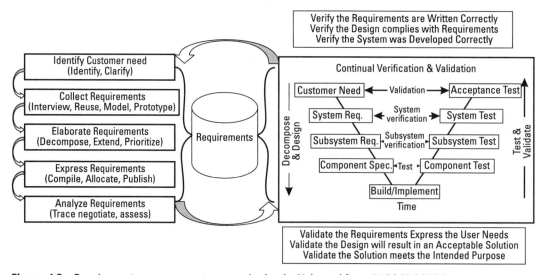

Figure 4.3   Requirements management process in depth. (Adopted from INCOSE 2017.)

Table 4.2   Requirements Terms

Requirement Type	Description
User needs	The problem or capability gap to be solved by the project.
Functional	What the solution must do (also known as initiating requirements).
Nonfunctional	Items required to support the system and its delivery (training, maintainability, sustainability, constructability, etc.).
System	Technical description of the system to be designed. Includes performance criteria for each element of the system.
Technical	A form of nonfunctional requirements that defines standards, codes, or practices that must be followed in the system design.
Specification	Design level requirements that define how the system is to be developed, constructed, and integrated.

shown in Figure 4.3, the requirements management process starts with defining the problem or need to be satisfied by the project. This need and other high-level project requirements are found within the project's initiating documents (scope of work, charter, or contract documents).

Step one is to review all initiating (or contract) documents to identify the goals, objectives, and expectations for the project. Project documents include

- Contract agreement;
- SOW;
- Terms and conditions;
- Contract data requirements list (CDRL);
- Supplier or subcontract data requirements list (SDRL);
- Data item descriptions (DID);
- Performance work statement;
- Statement of objectives.

Project objectives and outcomes will describe the problem or new capability the project is to deliver. These statements provide the basis for the project. Examples include

- The organization requires the ability to perform functions associated with a new line of business or mission;
- Currently the organization lacks the ability to perform specific functions;
- By increasing the range and lethality of its weapons, the organization will be able to destroy more targets at a greater distance.

Well-written project initiation documents will use clear "shall" statements to communicate the project expectations. These shall statements are the user needs (requirements) for the project. As with all requirements, these initial requirements statements shall be necessary, singular, achievable, unambiguous, and independent of a specific solution. Examples include

- The system shall_be capable of sensing a target at 100 miles;
- The system shall track a sensed target at speeds up to 100 miles per hour;
- The contractor shall deliver the first prototype within 100 days of contract award;
- The contractor shall provide the project earned value management report, as defined in section q, on the first of each month.

Not all customers can provide clear and concise shall statements. Often multiple needs or expectations will be combined into a single sentence. In other cases, the statements may lack the firmness provided by the word shall. Softer terms such as desired, should, may, and like are sometimes found in project documents. These terms are ambiguous and unmeasurable, nor do they support the validation methods that will be described later in this chapter. Additionally, subjective terms such as best practices, state of the art, military grade, durable, high performance, or optimum are sometimes used as performance descriptors for the system. These terms have the same issues because they leave too much open to interpretation for the customer and the project team. The project manager works with his or her team and the customer to clarify these ambiguous statements and formulate a set of clear and concise shall statements.

The INCOSE SE Handbook (INCOSE 2015); Business/Mission Needs Analysis Process provides amplifying guidance on how to identify and coalesce business or mission needs [2].

The project manager works closely with the project technical lead and the customer to clarify the customer's needs before moving on to the next step in the process. Missed or unclear requirements left unresolved at this stage will result in costly changes at later stages in the project.

Collect Requirements

The objective of the collect requirements process is to translate the user needs into a description of a set of solutions (not the solution) that will address the stated needs. Also known as requirements elicitation, the project team expands and refines the customer needs into a set of requirements that describe the functionality and high-level performance of the system.

There are many techniques and methodologies published on the topic of requirements elicitation and refinement. The sources listed at the beginning of this chapter provide a good reference to these techniques. Table 4.3 includes those methods most commonly used and recommended by INCOSE.

Interview technique involves structured communications with key stakeholders, such as clients, users, subject matter experts, and employees, to help clarify the requirements. The goal of this step is to clarify and refine existing requirements not to define a new set of requirements. The conversations and questions must focus on topics that will achieve this goal. Prepare for the interview by generating inter-

Table 4.3 Collection Methods

Method	Description
Interviews	Conduct structured interviews with project stakeholders.
Requirements reuse	Identifying existing requirements from legacy systems or similar projects that may apply to the current project.
Use case analysis	Modeling or describing user interactions with the proposed systems.
Task analysis	Modeling and or observing the organization's processes and procedures.
Prototyping	Development of system mock-ups, test beds, or simulations.

From: [3, 4]

view questions that will extract as much insight as possible to fully understand and clarify requirements.

A project manager and technical lead's role is to ensure that good interview questions are prepared, the appropriate interviewer is selected, the proper interviewee is selected, and the results of all interviews are documented. The person selected for the interviews requires a balance of technical understanding, requirements engineering experience, and the ability to communicate with multiple stakeholder types.

Examples of some of the key questions:

- What existing technology may be available for this product?
- Who is going to utilize this product?
- What factors are driving the project's budget, time, resource, and contract constraints?
- What requirements identified thus far do you consider to be the most important?
- What known performance issues exist with the existing product?
- Please provide specific temperature (hot, cold, or ambient) requirements?
- What are the atmospheric pressure requirements (high altitude or low altitude)?
- What specific facility requirements must be included in the design, development, and testing?
- What special tooling, equipment, and facility is needed to develop this product?
- Who else should we be contacting to help clarify these requirements?
- What requirements require further clarification from the customer?
- What are the key review milestones and what are the expectations to pass each review?
- What systems must this system interface?
- What specific training is required for the project team to successfully achieve project requirements?
- What is the security level of this project? Does it require clearances (secret, top secret, etc.) to work on the project?

- Who are the key suppliers?
- Does the technology for this project require an export license or nondisclosure agreement (NDA) or proprietary data sharing permission?
- What restrictions apply to outsourcing elements of this project?

Requirements reuse involves collecting requirements through existing procedures, requirements documents, user documents, regulations, and standards for similar systems or projects. Often projects involve improving or replacing an existing system for which its documentation can provide valuable insight into the functionality and performance required of the new system.

Use case analysis helps the users and designers to visualize the intended interaction between the user and the system. Each use case depicts a given function or interaction with the system by depicting the user's input as well as the response or output expected from the system. Use cases can be purely textual or a combination of diagrams and text. Requirements are refined and elaborated based on the observations gained through the use cases.

Task analysis involves documenting how the organization performs the functions to be supported by the new system. This approach is particularly helpful when the new system is intended to automate a manual process. Manual processes often lack standard procedures with each person doing things a little differently. Additionally, process documentation may not show every nuance of the steps required. By modeling the expected process then observing the users performing the tasks, any gaps in the documentation or inconsistencies in execution can be identified. Task analysis is also helpful in understanding potential interaction with systems external to the project. For example, during task analysis it is observed that after creating a record in the new system, the user needs some of this new data in a separate system. Even if this system-to-system exchange is already known, observing how the data is used provides clarity to the requirements for this integration.

Prototyping involves programming, modeling, samples, or initial circuit testing to demonstrate the requirements, resulting in a better understanding of the solution required. Often these mock-ups help the stakeholder to visualize the stated requirement and its impact on the system design. Prototypes can range from a simple white board diagram of the user functions, to screen mock-ups, to a functioning system, each of which provides varying degrees of functionality at equally varying costs. As the project manager, it is imperative to work closely with the project technical lead to understand the intended outcomes and investment required for prototyping exercises to ensure the work is supported by the project budget and leads to project completion.

While prototyping is a valuable method used by systems engineers, it must be used with caution. Prototypes can be confused with fully functioning solutions. Prototypes are only a demonstration of functionality in a controlled environment and are not always designed to operate under the stresses of the user's environment. Often the line between prototype and the final solution becomes blurred. In these cases, the prototype continues to expend effort to add more functionality, giving a false sense of capability to the customer. The resulting system is one that works if the developer is involved, not a usable solution.

There are usually three levels to measure product design: conceptual, development, and production.

The conceptual level means the design is based on trade studies and prototypes.

Development level means the product has received extensive development and testing.

The production level means the product can be built based on full demonstration and qualification of the product and its design [2].

All new requirements or changes to existing requirements identified during the collect requirements process shall be noted as to their source. Examples will be provided later in this chapter.

Throughout this step it is important to keep a good record of all interviews to clarify any misunderstandings or discrepancies after the fact. Keeping minutes of the meetings and submitting the records to those involved to receive confirmation of their accuracy is also recommended.

Communications between technical and nontechnical people can be a challenge. Using complex jargon when speaking to the nontechnical person may be intimidating and an interviewer that does not understand the technology being discussed could result in lost details. Both cases may result in eroded confidence in the project team. One of the key roles of the project manager during interviews is to help the interviewee feel comfortable, to make sure they are not being challenged, and ensure that the goals and objectives of the meeting are being met. Sometimes it takes a third party to identify potential missed communications between two other parties.

The most important thing in communication is to hear what isn't being said.
—Peter Drucker

Requirements collection and elicitation requires strong communications and analytic skills as well as a firm understanding of the user's business/mission, and the environment in which they execute their mission. Successful project managers are those that can identify the knowledge and skills needed and can facilitate collaborative relationships between those involved in the requirements process.

Elaborate Requirements

The objective of the elaborate requirements process is to decompose and refine the previously documented user need statements and functional requirements into a technical description of the solution (specifications). Elaboration involves the iterative decomposition of each requirement down to its lowest level of detail.

Requirements elaboration involves the activities depicted in Table 4.4.

Decompose. Requirements at project initiation are likely to be high level and will evolve over time. As with scope and schedule management, requirements management requires disciplined configuration, and change management to ensure that new discoveries do not become unapproved scope. As with the project scope discussion, the goal is to increase fidelity of the requirements while controlling scope.

Table 4.4 Requirements Elaboration Activities

Elaborate Activities	
Activity	*Description*
Decompose requirements	Translate each functional requirement into a system attribute that will satisfy the function. Translate the functional and nonfunctional requirements into design specifications.
Extend: identify technical requirements	Assign technical/regulatory requirements to each requirement or group of requirements.
Extend: define nonfunctional requirements	Identify elements required to support or deliver the solution.
Prioritize requirements	Assign weights to each requirement.

When new requirements are identified the impact on the project scope is managed by following the techniques described in Chapters 2 and 3.

Extend. As requirements are decomposed into lower-level specifications, be sure to engage the project stakeholders that have technical and regulatory voices in the project. During this activity, nonfunctional and technical requirements will be the predominant type identified. Within Chapter 3, an example was provided where the project team learned at preproduction the system design did not include a key safety standard. This previously unidentified technical requirement created major impacts on the project's cost and schedule.

Unfortunately, technical requirements are not often found within a single source. The PM's role in this process is to work with the project technical lead to ensure all appropriate sources and stakeholders are identified and engaged. Often those with the most expertise are also the ones with the least availability. These stakeholders may prioritize their engagements based on the level of risk associated with each project in its life cycle. These experts will give priority to a mature project design approaching production approval over a project in the requirements definition phase. However, waiting until later in the project life cycle to acquire subject matter experts (SME) engagement will expose the project to unacceptable risk.

> *Technical and Regulatory Stakeholders are those with explicit or implied authority to impose requirements on projects. Technical Authority is assigned to independent representatives of the customer organization or jurisdictions in which the project resides. Technical authorities are charged with the responsibility of enforcing standards for safety, interoperability, quality, security or processes.*

The project's initiating documents (statement of work, charter, contract, solicitation, etc.) will contain references to standards in which the project shall comply. Do not assume that every document listed has a clear applicability to the project. A common complaint among project managers and engineers is the cover-all-bases approach that many contract writers follow. In this approach, a multitude of required references are added to the document to cover missed details on the part of the customer. This approach levies unnecessary requirements on the project and exposes the customer and the project team to potential risk due to this lack of clarity in the scope.

Often these documents are listed within the project documents without specifying which parts of the documents are important or how the documents are to be applied to the design. This writer has experienced multiple cases where a relatively small contract of 100 pages referenced more than 10,000 pages of supporting material. It's not realistic to read all the material, but it is very important to be familiar with the references and focus on those parts that directly impact the project. Rely on project technical experts to help qualify the relevant documents. This is a case where asking the technical representative a specific question, "What section or sections of ISO 24051:2010 do you follow within your area of technical authority," may result in quick response rather than asking them an open-ended question about their requirements. It may seem counterintuitive to go looking for requirements but the potential impact on the project for failing to identify unrealistic technical requirements is real.

Lesson Learned, Read All the Documents
The housing on a radar system for the F-15 was a cast component. Embedded in one of the 68 secondary documents was a material restriction that specified the minimum elongation (a measure of how brittle the material is) before failure. The specification also stated that if elongation value was less than a second, slightly higher value Marine Corps Air Command (MCAIR) approval was required. This statement basically set the customer's expected performance at the higher value.

The drawing for the casing had been updated many years ago and the only elongation value depicted was that of the minimum value. All purchased castings were under customer expected value and some were exactly at the minimum. The entire batch of supplied housings were out of compliance with the spec as no MCAIR approval had been requested. The lead time to get new castings was very long. The only reason this oversight was caught was the customer had requested another minor assembly change which drove the analysis team to re-look at the casing design. Without this catalyst, no one would have ever identified the noncompliance. A waiver was requested and granted from MCAIR to use the castings as-is until detailed analysis was conducted to show this noncompliance would not be detrimental to the systems performance. This oversight cost tens of thousands of dollars in additional engineering, meetings, and paperwork. Reading specifications and referenced documents early in the program will help to prevent these types of mistakes [7].

When identifying lower-level specifications, it is important to read through secondary reference documents to decompose accurate requirements for the component level. It is important that the source of all new requirements be noted. Each new technical requirement shall reference the document and section in which the requirements were identified.

Prioritizing requirements. One of the biggest reasons to prioritize requirements is because there are usually more requirements than resources to achieve the requirements. As a result, low-priority requirements may not be immediately implemented due to time and budget constraints. Prioritizing requirements early in the project will help to guide decisions later. Customers tend to believe that prioritizing means they may be forced to give up a feature or function later. A requirement is a validated need and prioritization should not imply that something is nice to have or desired, which implies lack of validity. It can be helpful to use priorities that do not reduce the validity of the requirements. Two such methods for prioritization are the specificity and timing. In specificity, the customer is acknowledging the potential to accept alternative solutions and in timing the customer is accepting potential phasing of the functionality. The challenge is to get stakeholders to admit all requirements are not critical. Table 4.5 shows an example of the priority levels for these techniques.

In either case, the requirements are being prioritized based on how or when they are delivered while maintaining validity for the requirement.

The elaboration process continues until the requirements provide the specificity required to design and produce the system. Techniques such as modeling and prototyping help the customer and the project team visualize the requirements and to identify any potential inconsistencies and ambiguities.

Express Requirements

The objective of express requirements is to establish a single data set of the requirements statements that can be presented, communicated, understood, and utilized in project completion. The activities performed to express requirements are depicted in Table 4.6.

Create requirements data set. Collecting, organizing, and managing requirements without a specialized tool can be a challenge. Spreadsheets can provide tabular presentation structure often used to display and share requirements. However, these tools may present challenges when the data set is large and the relationships to upstream and downstream documents are numerous. The objective is to establish the capability to create a single data set in which all requirements are stored, shared, assessed, managed, and decomposed. Requirements are tracked and managed using data attributes or tags to denote status, types, sources, and linkages to design and testing documents. Many requirements tools provide these attributes by default and support the addition of custom fields.

Table 4.5 Example Prioritization Methods

Priority	Specificity	Timing
High	Must be exactly as specified	Must be included in first increment
Medium	Must perform as specified	May be able to shift to second increment
Low	Alternative solutions will be considered	May be able to shift to a third increment

Table 4.6 Express Requirements Activities

Organize Requirements	
Activity	Description
Initiate requirements data set	Select requirements management tool and configure requirements attributes.
Allocate requirements	Put requirements into logical groups and begin to assign them to elements of the WBS.
Present/publish	Create reports, diagrams, or presentations

Allocate requirements. Up to this point, requirements may not be grouped into specific categories such as subsystems or functionality. There may be duplicative or similar requirements undetected until they are grouped together. The objective of this step is to begin to coalesce the requirements so that they can be codified and then linked to their sources and downstream artifacts. The systems engineer or requirements manager will use an array of techniques to accomplish this task, logging the results in the requirements database using attributes or data tags.

Publish requirements. Requirements must be read, understood, and agreed upon to be relevant to the project outcomes. Historically, requirements were published in large documents and matrixes. Table 4.7 is an example of a preliminary requirements matrix. This example was produced using a professional requirements

Table 4.7 Example Requirements Matrix

Source	Number	Name	Description	Creation Date
	1	Q36 space modulator requirements		8/7/2018
Project SOW and charter	1.1	Problem statement	New threats require an improved capability	8/7/2018
Project SOW and charter	1.1.1	Sense targets	Ability to sense targets	8/7/2018
Project SOW and charter	1.1.1.1	Sense range	The system shall sense targets within 100 miles	8/7/2018
Interview with operations	1.1.1.1.1	Range of 100 miles	The system shall sense a target at a distance up to 100 miles	8/7/2018
Interview with operations	1.1.1.1.2	Target environment	The system shall sense a target in space atmosphere	8/7/2018
Project SOW and charter	1.1.1.1.3	Target sense speed	The system shall sense targets moving between 1 mile and 900 miles per hour	8/7/2018
Project SOW and charter	1.1.2	Destroy targets	Â ability to destroy targets	8/7/2018
Interview with operations	1.1.2.1	Engagement range	The system shall engage target at a range of Â 75 miles	8/7/2018
Interview with operations	1.1.2.2	Engagement speed	The system shall traverse distance between shooter and target before target can reach shooter	8/7/2018
Project SOW and charter	1.1.3	Track targets	Ability to track targets	8/7/2018
Project SOW and charter	1.1.3.1	Target tracking	The tracking system shall maintain a track of target movements	8/7/2018
Project SOW and charter	1.1.3.1.1	Tracking speed	The system shall track targets moving up to 900 miles/hour	8/7/2018
Interview with operations	1.1.4	Identify targets	Ability to identify targets	8/7/2018
Interview with operations	1.1.4.1	Target identification	The system shall match the target against a set of known targets	8/7/2018
Interview with training	1.1.5	Target discrimination	Ability to discriminate between enemy and friendly targets	8/7/2018
Interview with training	1.1.5.1	Designate enemy targets	The system shall designate enemy targets	8/7/2018
Interview with training	1.1.5.2	Designate enemy targets	The system shall designate friendly targets	8/7/2018

management (RM) tool preconfigured with the common data attributes associated with requirements management. The requirements shown are now grouped into functional elements with the functional requirements as the parent and the children defining the system's performance requirements, all of which link up to the initial problem statement and capability statements.

While these tabular presentation methods remain useful, graphical approaches have added a perspective to requirements sets often missed in text-based formats. One example of a graphic presentation technique is the requirements model. Requirement models use a hierarchical structure like the work breakdown structure. Also known as a product breakdown structure, specification tree, or a requirements breakdown structure, requirement models visually demonstrate relationships between requirements with parent-child relationships and dependencies. The method helps the reader to see how the satisfaction of higher-level requirements is predicated on the satisfaction of all lower-level requirements. An example requirements breakdown structure is shown in Figure 4.4. Table 4.7 and Figure 4.4 were created from the same RM tool and database. By using this tool, two views of the same data can be published.

By using professional tools to create, manage, and present the requirements data, the project team will be positioned to move the requirements to an acceptable level of maturity and receive meaningful engagement with reviewers.

Analyze Requirements

The goal of analyze requirements is to determine the integrity of the requirements data set and resolve any potential gaps or issues. As requirements come from multiple sources, there may be occasion that two similar requirements contain conflicting information. In other cases, there may be requirements with unresolved questions. The PM's role in this process is to work closely with the project technical lead to track the progression of issue resolution and to facilitate the resolution of stalled items. Table 4.8 provides a list of common techniques and activities associated with analyzing requirements.

Verification methods: The words verification and validation are used throughout project management and systems engineering with each case having different applications and relevance to the project life cycle. For this discussion the following is adopted from [3].

Verification is the process to confirm that a requirement, design, or system is correct against set criteria, standards or practices.

- Requirements are verified to be consistent with established practices and syntax (For example, *INCOSE Guide to Writing Requirements*);
- System designs are verified to be compliant with established practices and the project requirements;
- The system is verified to have been developed according to established practices, technical standards, and the design.

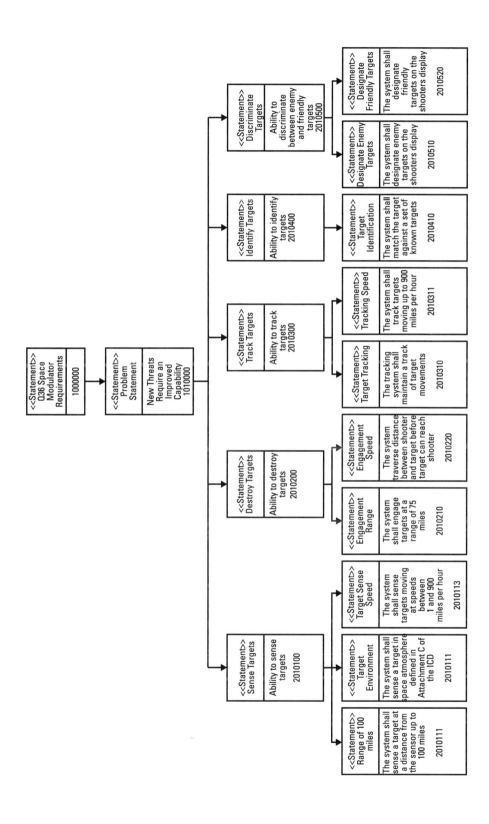

Figure 4.4 Example of requirements model.

Table 4.8   Requirements Analysis Techniques

Analyze Requirements	
Activity	*Description*
Define system verification criteria	Assign a system verification method to each requirement
Trace requirements	Ensure each requirement traces to a parent or source. Ensure each requirement traces to the system design.
Negotiate	Collaborate with stakeholders to achieve resolution to open issues with requirements
Assess requirements	Assess maturity and completeness of requirement statements

Validation is the process to confirm that a requirement, design, or system is consistent with its preceding descriptions.

- Requirements are validated to confirm the requirements accurately express the user needs;
- System designs are validated to confirm the design will result in an acceptable solution;
- The system is validated to confirm the solution meets the intended purpose of the project.

In this step the systems engineering team assigns the proposed method by which the design and the resulting system will be verified. The most common verification methods are shown in Table 4.9.

Each of these methods has advantages and disadvantages based on the requirement and system element to be verified. The goal of this step is to ensure the project deliverable is consistent with the stated scope and requirements. The customer is verifying they receive what they bought, and the developer is verifying that they deliver only what was contracted. A more costly or complex verification method may protect the customer in some case and the developer in other cases. Defining and agreeing on verification criteria early in the project establishes expectations for acceptance and closure of each product deliverable. These acceptance criteria are also useful when determining liability for issues with the system after acceptance. The PM as the accountable party for project success (cost, schedule, and performance) has a critical interest in the outcome of this process.

Table 4.9   Verification Methods

Requirements Verification Methods (A, D, I, S, and T)	
Analysis	A quantitative of a complete system or subsystem by review and analysis of collected data.
Demonstration	To prove or show, usually without measurement of instrumentation, that the product or syustem complies with the reuiremnts by observation.
Inspection	To examine visually or usesimple physical measurement techniques to verify conformance with specified requirments.
Similarity	To show capability of functioning based on knowledege for similar components or materials.
Test	A measurement to prove or show, usually with precession measurements or instrumentation, that the product or systems complies with requirements.

Trace requirements. Requirements are the single thread that goes through a project life cycle from its conception through design, build, test, and customer acceptance. Tracing establishes the thread from the user needs through the requirements and into the system design, delivery, and testing. Requirements are traced

- To the originating problem or needs statement;
- To the originating document that contained the requirements or need statement;
- To meetings or discussions where requirements were defined, communicated, or approved;
- Down to child requirements and up to parent requirements;
- To allocation to the WBS, system elements, subsystems and components;
- To the design documents that translate the requirement into the design;
- To the verification methods planned for each requirement or requirement set.

Tracing is logged within the requirements database using tags or attributes for each trace relationship. Table 4.10 was published from the same requirements database as the prior tables and figures with the additional trace attributes now assigned to each requirement.

A requirement that cannot be traced to a parent or a source document is referred to as an orphan and may indicate a change in scope or scope creep. A requirement that is not demonstrated within the design documents, allocations, and test cases is an unsatisfied requirement. In both cases the single thread through the project is not demonstrated.

Negotiate requirements. During evaluation of the requirements there might be cases where requirements cannot be listed as achievable or feasible due to technology, cost, or schedule constraints. In other cases, technical standards and safety regulations may preclude the implementation of certain requirements. Therefore, the customer may decide to deviate (provide relief) from the requirement if the engineering analysis proves it does not impact overall performance of the system. When deviating from requirements make sure it does not impact form, fit, or function of the product. Although requirements can sometimes be waived, customer needs or key functional requirements are rarely waived as this would impact the efficacy of the project.

The following are two examples of where requirement relief was granted for two different fundamental reasons.

On a $300M+ NASA satellite project, the engineering requirements document (ERD) was rushed together with limited systems engineering support early in the project definition. The most conservative assumptions were made at each decision point and this led to a very severe specified operating environment. Initial design and analysis showed that a redesign the structure would be required to accommodate the high loads. This change would have tremendous cost and schedule impacts. The requirements team was able to review their environmental data assumptions and make more reasonable (and less conservative) predictions for the expected loads. Some critical load cases were reduced by a factor of four. The system no longer needed redesign and would still meet the requirements for safe use

Table 4.10 Example Requirements Trace Results

WBS	Source	Number	Name	Description	Creation Date	Modify Date	Design Document	Verification Method	Allocation	Test Case
	Project SOW and charter	1		Q36 space modulator requirements	8/7/2018	3/7/2019				
	Project SOW and charter	1.1	Problem statement	New threats require an improved capability	8/7/2018	3/8/2019	NA			
2.00	Project SOW and charter	1.1.1	Sense targets	Ability to sense targets	8/7/2018	3/9/2019	NA	D	Targeting system	TC22
2.00	Project SOW and charter	1.1.1.1	Sense range	The system shall sense targets within 100 miles	8/7/2018	3/10/2019	System Design Spec (SDS) P22,	A	Targeting system	TC22
2.00	Interview with operation	1.1.1.1.1	Range of 100 miles	The system shall sense target at a distance up to 100 miles	8/7/2018	3/11/2019	SDS P22,	A	Targeting system	TC22
2.00	Interview with operations	1.1.1.1.2	Target environment	The system shall sense a in space atmosphere	8/7/2018	3/12/2019	SDS P12,	A	Targeting system	TC22, TC26, TC28
2.00	Project SOW and charter	1.1.1.3	Target sense speed	The system shall sense targets moving between 1 miles per and 900 miles per hour	8/7/2018	3/13/2019	SDS P32	I	Targeting system	TC26
4.00	Project SOW and charter	1.1.2	Destroy targets	Ability to destroy targets	8/7/2018	3/14/2019	SDS P32	D	Weapon payload	TC28
4.00	Interview with operations	1.1.2.1	Engagement range	The system shall engage target at a range of Å 75 miles.	8/7/2018	3/15/2019	SDS P33	T	Weapon payload	TC28
4.00	Interview with operations	1.1.2.2	Engagement speed	The system shall traverse distance between shooter and target before target can reach shooter	8/7/2018	3/16/2019	SDS P34	T	Weapon payload	TC28
3.00	Project SOW and charter	1.1.3	Track targets	Ability to track targets	8/7/2018	3/17/2019	SDS P45	A	Targeting system	TC15
3.00	Project SOW and charter	1.1.3.1	Target tacking	The tracking system shall maintain a track of target movements	8/7/2018	3/18/2019	SDS P46	A	Targeting system	TC15
3.00	Project SOW and charter	1.1.3.1.1	Tacking speed	The system shall track targets moving up to 900 miles/hour	8/7/2018	3/19/2019	SDS P47	A	Targeting system	TC15
2.00	Interview with operations	1.1.4	Identify targets	Ability to identify targets	8/7/2018	3/20/2019	SDS P48	T	Targeting system	TC15

Table 4.10 (continued)

WBS	Source	Number	Name	Description	Creation Date	Modify Date	Design Document	Verification Method	Allocation	Test Case
5.01	Interview with operations	1.1.4.1	Target identification	The system shall match the target against a set of known targets	8/7/2018	3/21/2019	SDS P49	T	Targeting system	TC22
5.02	Interview with training	1.1.5	Target discrimination	Ability to discriminate between enemy and friendly targets	8/7/2018	3/22/2019	SDS P50	I	Targeting system weapon safety subsystem	TC11
5.03	Interview with training	1.1.5.1	Designate enemy target	The system shall designate enemy targets	8/7/2018	3/23/2019	SDS P51	I	Targeting system weapon safety subsystem	TC11
5.03	Interview with training	1.1.5.2	Designate enemy target	The system shall designate enemy targets	8/7/2018	3/24/2019	SDS P51	T	Discrimination subsystem	TC11

in a launch environment. At this later stage in the project, more information was available to supported more accurate design estimations. Collaboration allowed for the successful project completion.

On another NASA project, a docking adapter specification included a not to exceed dimensional tolerance. The specification was extremely tight, and two engineers each spent a year trying to get the design to meet this requirement. In the third year in of design, a new engineer implemented a multi-variable optimization of the structure which indicated all potential solutions were outside the available design space. In other words, there was no feasible or obtainable solution to the given the requirements and design constraints. It was an impossible problem. When this overwhelming amount of support data was provided to NASA, they realized that yes, in fact the requirements were impossible to meet, and, in fact, the other vendors were not meeting the requirements either. The tolerances for the requirement were changed to reflect what the optimization model depicted as the achievable space [7].

In both examples it is hard to say that the project team should have identified the problem early in the project. However, in both cases the requirements not listed as achievable or realistic should have indicated project risk. Identifying risks associated with these requirements should have elevated their visibility, increasing the chance the issues would be identified earlier in the project.

Requirements Verification and Validation

As discussed in the previous section verification and validation is performed at multiple stages in the project. As this chapter is focused on requirements management, verification and validation here applies to activities associated with requirements.

Requirements verification is the process to confirm that each requirement is consistent with established practices and syntax [3].

Requirements validation is the process to confirm the requirements set accurately expresses the user needs [3].

Verify requirements. Within industry publications there are several methods for evaluating the maturity or completeness of a requirements. These methods include a set of attributes in which all shall be met for the requirement to be considered mature. The acronym SMART is used in many practices to define a quality requirement, objective, or goal. However, numerous variations exist on the words that compose SMART. While using a variation of SMART is acceptable, the metrics published by ISO and INCOSE are much more consistently applied and used herein.

- *ISO 15288:2015 and INCOSE SE Handbook*: Defines a requirement as unambiguous, clear, solution independent, unique, consistent, stand-alone, verifiable, and necessary [1,2];
- *INCOSE Guide to Writing Requirements* provides extensive criteria for the formulation of individual and sets of requirements [3].

They key point to this discussion is that requirements regardless of the type, (functional, nonfunctional, technical, or specification) must be evaluated against an agreed upon set of criteria to gauge the maturity and completeness of the requirements. Table 4.11 includes an aggregated set of the common quality measurement attributes.

The PM's role in this process is to work closely with the project technical lead to monitor the requirements process using objective characteristics such as those provided to gauge the overall maturity of the requirements. As the project schedule is the instrument used to measure overall project progress, the requirements verification process is the instrument to measure the completeness of the requirements.

Requirements verification is enabled by following the processes and techniques described up to this point. The objective of requirements validation is to confirm the requirements accurately express the user needs. Verification of this accuracy is demonstrated through the following:

- Ratification of the customer needs (problem statement and functional requirements) establishes agreement on the reason for the project.

- Requirements that can be traced up to its source and each user need statement can be traced down to its elaborated requirements. Need statements are not fully expressed until they are traced down to the specification level.

- Distributing and receiving concurrence on the results of requirements meetings demonstrates the communication of requirements.

- Tracking the person and event in which requirements were communicated demonstrates the user's voice in the process.

- All potential issues with requirements are address early in the project reducing ambiguity.

- Requirements are reviewed in a structured process in which all stakeholders have been afforded adequate input.

- Requirements identified and traced to through detailed analysis of the reference documents helps to ensure input from technical stakeholders.

Table 4.11   Requirements Quality Measures

Attribute	Description
Clear	Worded using accepted grammar and is unambiguous or open to interpretations.
Complete	Communicates complete thought using recommended requirements diction (shall statements, no soft words or measures).
Unique	Communicates a single statement.
Consistent	Does not conflict with other requirements.
Correct	Describes the stakeholder's intent and is validated through a review process.
Design	Does not impose a specific solution. Describes that the product, system, or project must do, not how to do it.
Feasible	Implementable with existing or projected technology and within cost and schedule.
Traceable	Uniquely identifiable and able to be tracked to its sources and downstream design and development entities.
Verifiable	A realistic means to verify the requirement is being met within the design and development.

From: [2, 3]

- Objective criteria used to validate each requirement to determine its maturity and completeness. Several of the characteristics in Table 4.7 directly support the verification that the requirements process has captured and translated the user's needs into a set of design criteria.

The PM and project technical lead's role in the verification process is to ensure that the activities and techniques described in this chapter are followed, which will result in verification that the user's needs are accurately expressed.

Summary

This chapter presented the concepts of requirements management as an important activity required for successful project management.

Throughout the requirements management process the requirements are traced from their higher-level need statements down through the system design. Requirements verification is used to measure the quality and maturity of requirements according to objective criteria. Requirements validation confirms that the requirements accurately describe a solution to meet the user's needs and expectations.

References

[1]   International Organization for Standards (ISO), International Electrotechnical Committee (IEC) and Institute of Electrical and Electronic Engineers (IEEE), "15288:2015 Standard for Systems and Software Engineering–System Life Cycle Processes," International Organization for Standards (ISO), Geneva, 2015.

[2]   International Council on Systems Engineering, *Systems Engineering Handbook: A Guide For System Life Cycle Processes and Activities*, Hoboken, NJ: John Wiley & Sons, 2015.

[3]   International Council on Systems Engineering, *Guide for Writing Requirements*, San Diego, CA: International Council on Systems Engineering, 2017.

[4]   Fostburg, K., H. Mooz, and H. Cotterman, *Visualizing Project Management: Models Frameworks for Mastering Complex Systems*, Third Edition, Boston: John Wiley & Sons, 2005.

[5]   Project Management Institute, *Pulse of the Profession* (2014), Newtown Square, PA: Project Management Institute, 2014.

[6]   Jones, C. M., Director, *Haredevil Hare* [film], United States: Warner Brothers Cartoons and Vitaphone, 1948.

CHAPTER 5

Risk Management

Risk is any uncertain event or condition that if it occurs will influence one or more project objectives. These influences on a project can be opportunities (positive) or threats (negative). Risks are any future uncertainties in achieving the objectives of a project within its cost, schedule, and performance constraints. This chapter provides the technical project manager with an introduction to the terms, methods, and information elements of risk management. The risk management process is an example of an organizational process asset that is owned by the PM's organization and PM tailors the process to the project's specific needs to capture project risks throughout the project life cycle. Typically, a project manager will spend at least 30%–40% of his or her time upfront generating the project risk management plan. Throughout the project, the project manager may spend 5–10% of his or her time managing risks on monthly or as required basis. Project managers are encouraged to seek the assistance of qualified risk management practitioners and to be familiar with the following reference which is cited throughout this chapter.

Risk Management Guide for DoD Acquisition, Sixth Edition (Version 1.0). The U.S. Department of Defense, Office of the Under Secretary of Acquisition, Training and Logistics (OUSD-AT&L) [1].

This guide contains baseline information and explanations for a well-structured risk management program. The management concepts and ideas presented in this document are useful in adopting risk-based management practices and suggest a process to address program risks without prescribing specific methods or tools.

Risk management is inherently proactive in nature as its goal is to identify conditions that may impact the project and it requires planning to address these conditions in advance of their occurrence. Risks are events that have not yet occurred. Issues are risks that have occurred and are now impediments to achieving the project's objectives. Risk management does not eliminate the occurrence of all risks, but it can reduce the severity of their impact if realized. Without planning, the chances of minimizing the scale of these impediments is greatly reduced. The objective of risk management is to realize the benefits of opportunities while minimizing the impact of threats. The risk management process involves the planning, identification, analysis, monitoring, and response of project risk [1].

Risk can exist at the project level, and can be associated with a project objective, a single deliverable, or work package within the project. Figure 5.1 depicts the core activities associated with risk management. Note that risk management

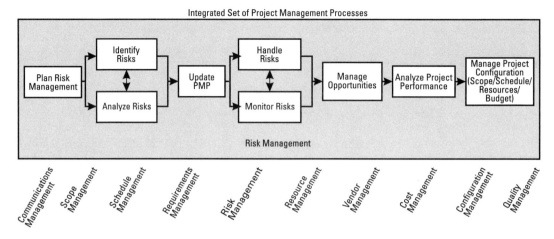

Figure 5.1   Risk management process.

planning is performed once or at predefined intervals, but the remaining activities are repeated whenever change is planned, change occurs, and at predefined intervals [2].

Risk management is tightly coupled to the other project management processes because risk can impact the cost, schedule, or performance associated with the project outcomes as shown in Figure 5.2. As discussed in the previous chapters, a change to any of these elements will likely result in a change to the other two.

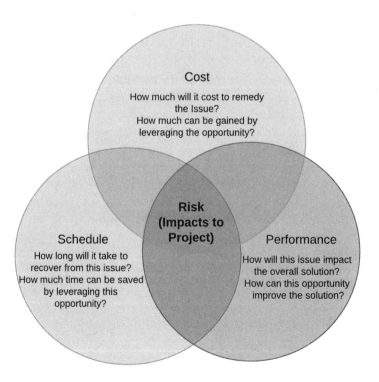

Figure 5.2   Impact of risk on cost, schedule, and performance.

Cheap or expensive missions can both succeed, but rushed missions almost always fail. Failure is the norm: something goes wrong on almost every project.
—*Space Systems Failure* by David M. Harland and Ralph D. Lorenz [3].

The project manager leads and enables the risk management process, relying on the expertise and authorities of those assigned to the project team and within his or her organization. Table 5.1 defines the roles and responsibilities associated with the processes described within this chapter.

Table 5.1   Roles within Risk Management

Role	Responsibility
Customer lead	The member of customer's organization charged to manage the project from the customer's perspective. Has final authority over all risk responses that may involve additions, changes, or deletions to project scope, schedule, and performance.
Senior leadership: Project organization	Members of the project manager's organization with the accountability and authority to apply the organization's resources to risk response and to accept risk exposure to the organization. Project success or failure exposes the organization to positive or negative outcomes.
Project manager	The person accountable for project success, the PM leads, monitors, and supports the risk management processes and procedures.
	Encourages senior leadership involvement in risk management.
	Works with senior leadership and stakeholders to determines acceptable risk levels.
	Openly and proactively communicates risk internally and externally to the team.
	Escalates risk to the customer and senior leadership in cases where predetermined thresholds have been reached.
Procurement officer	Member of the PM's team that can authorize new agreements and changes to agreements between the project team and the customer or between the project team and its vendors.
Project stakeholders	Each project stakeholder's risk tolerance, sensitivity, and motivations should be documented as part of the stakeholder identification and management process.
Subject matter experts	Members of the PM's organization who are engineers, scientist, and analysts with expertise in one or more area of the project. The PM relies on the experiences possessed by these experts.
Risk manager	Member of the PM's team with training and experience in risk management. The risk manager provides guidance in the identification and analysis of project risks then manages the consolidation, structuring, publication, and management of the project risk plans and data.
Work package lead	The person assigned responsibility and accountability to deliver a work package. May be a member of the PM's organization or the contracted organization if the work package is outsourced. The work package lead provides risk input associated with the individual work package.
Project scheduler	Member of the PM's team that performs the consolidation, structuring, publication, and management of the project schedule. The scheduler works closely with the risk manager to link risks to the project WBS and schedule.
Risk working group	A subgroup of the project team charged with initial planning for risk management and the continual identification, analysis, monitoring, and response to risks.
Risk management board	May exist on larger projects. Comprised of representatives of leadership and the customer organization. This group provides risk management awareness and accountability while assisting in the selection of risk responses (mitigate, transfer, avoid, or accept). This group has the authority to commit resources and management reserve to handle risks and issues.

A seasoned risk manager shares an experience obtained while working on a project for the U.S. Air Force. During a risk planning meeting, the project customer, a U.S. Air Force Colonel, made the following statement: "Risk is simply identifying all the problems that could occur, analyze how to prevent them, and planning how to resolve them, before they become an issue. We want to be in a proactive mindset, and not a reactive one." The Colonel proceeded to share the following example: "If I'm going to the gas station to fill my car up, what could possibly go wrong?" [4]

In this discussion the Colonel used an everyday activity to communicate the importance of risk management. This use case is explored in the following sections to demonstrate the risk management process.

Plan Risk Management

Plan risk management is the process used to describe how the project team will perform the identification, analysis, monitoring, and response to risks. The objective of risk management planning is to establish the project's risk management strategy and the organizational elements required to ensure risk management is an integral element of the project's management structure. PMs should avoid the temptation to reinvent the wheel; usually the PM's organization owns the risk management plan and has identified the level of risk management rigor expected based on the contract value of the project. A PM should expect that with higher contract values, a higher level of risk management rigor will be expected by his or her organization.

The output of this step is a risk management plan that documents the process of risk identification, analysis, handling, and risk monitoring, along with the roles and responsibilities of the risk organization. Mature project management organizations have standard project management processes, templates and guides that should include risk management. These templates provide the organization's standard risk management processes, tools, and reporting formats that may be adopted.

The risk management plan should include the following information elements:

Introduction. A brief description of the use and purpose of this document along with the intended audience and the organizational process assets used to develop the risk management plan.

Relationship to the project management plan. If this is a stand-alone document, describe how this document relates to the project management plan. Avoid duplicating detailed information provided within the project management plan.

Project description. Provide a high-level description of the project and its objectives. During project initiation, a set of measurable project objectives and success criteria were defined. It is imperative that the project objectives be accurately defined so that risk management activities address all risks that have potential to impact the project. This concept will be demonstrated in Section 5.2.

Risk management overview. Describe the risk management methodology and how this methodology aligns to organizational process assets or industry standards. Assuming an organizational standard process exists, the team determines if the processes require any adjustments or tailoring for the specific project. Reasons for tailoring may include level of exposure to the organization, political attention, complexity, or experience with the problem space. If no process or templates are available, a trained risk manager may start with the *Risk Management Guide for DoD Acquisition* referenced in the beginning of this chapter.

Ground rules and assumptions. These provide the reader the basis for the decisions made during the planning process. The assumptions may include the contract threshold, urgency to deliver, acceptable risk levels, risk exposure to the organization, political attention, or project complexity. Ground rules will include the frequency at which risks are reassessed, triggers for escalation, reporting requirements, expectations for all team members, and rules for the allocation of budget and schedule reserves.

Risk likelihood and consequence scoring criteria. Describe the criteria to be used for assessing the likelihood and consequence of risk occurrence used during risk analysis. These data points are demonstrated in Section 5.3.

Risk management organization. Describe the risk management organization, along with any risk management working groups and the roles, authorities, and responsibilities of all involved in risk management. Include any updates to the stakeholder matrix that defines stakeholder risk tolerances.

The RACI matrix is a good tool to use when communicating the roles and responsibilities of risk management. This approach encourages the team to think about their individual role in risk management instead of assuming all work will fall on a single person. Figure 5.3 is an example of a RACI matrix in which the relationships associated with each activity are:_

- (R) *Responsible*: the person who is assigned to work on a task;
- (A) *Accountable:* the single person who is authorized to make decisions and take ownership of a task;
- (C) *Consulted:* one or more people responsible to contribute on the task;
- (I) *Informed:* anyone who requires awareness of the task or individual risks.

The risk management process. This section describes each process that will be used for risk management. Process documentation should include both a process graphic and narrative describing the process along with expected inputs, outputs, and performers. Assuming there exists organizational standard process documentation, the risk management plan may simply reference the existing documents. However, the team may determine the standard processes requires adjustment or tailoring for the specific project, the tailored processes are documented here. This section of the plan should also include the standard format to be used for all risk statements.

RACI MATRIX	Project Manager	Project Team	Risk Owner	Mitigation Plan Owner	Program Manager/Sponsor
Risk Planning	A	I	C	R	C
Risk Identification	A	R	R	C	I
Risk Analysis	A	I	R	R	
Risk Handling	A	I	R	R	I
Risk Monitoring & Control	A	I	R	C	C

Figure 5.3 Risk management RACI matrix.

If no process or templates are available, a trained risk manager may start with the references listed in the beginning of this chapter. Table 5.2 provides a list of the subprocesses associated with risk management described in this chapter along with the activities that should be addressed in each subprocess.

Tools and templates. Document any risk management and analysis tools that will be used during the process. When selecting tools, be sure to identify adequate licenses and supporting infrastructure required by the tools.

Risk management is the formal process of identifying, analyzing, and monitoring risks. Risk is an unplanned event or condition that can impact the overall project, a project objective, or an element of the project. Risk management is a specialty process knowledge area that requires specific skills and experience. Potential impacts to a project include cost, schedule, and performance. Issues are risks that have been realized and are now impacting the ability to meet project objectives. Risk management plans document the results of the planning process, which should be supported by all project stakeholders and subject matter experts.

Risk and issue management are similar processes: risk is an unplanned event and should be handled appropriately before becoming an issue, an issue is a known event that has or is expected to occur. Issue management involves the reactive allocation of resources to address the impacts of impediments to project success. Risk management involves the proactive application of resources to mitigate the occurrence and impact of issues before they occur.

Table 5.2   Risk Process and Activities

Subprocess	Output Task	Activities
Plan	Risk management plan	Plan risk management: establish risk processes, procedures and orgainzation
Identify	1. Root cause 2. If…Then statement	Identify project risks and assign to responsible owners
Analyze	1. Likelihood and consequence 2. Ranking and prioritizing	Perform qualitative analysis on individual risks to establish probablity, impact, root causes, and prioritization.
		Perform qualitative analysis on the collective risks to evaluate overall project risk exposure.
Handle	Decide implementation approach	Plan risk response strategies. Define response actions, action owners and timlines. Execute and manage the risk response plans and strategies.
Monitor	Document traceability	Monitor risks and track status, lessons learned and risk patterns.
Communicate	Improve the project	Communate risk, risk status and influenece engagement in risk management.

From: [1]

This information can be used by the project manager to help gauge the topics that should be covered within the risk management processes section of the risk management plan. Figure 5.4 depicts these subprocesses and how each relates to the other subprocesses. Each subprocess is supported by activities that result in expected inputs/outputs. Across the bottom of the diagram is a list of recommended participants in each subprocess.

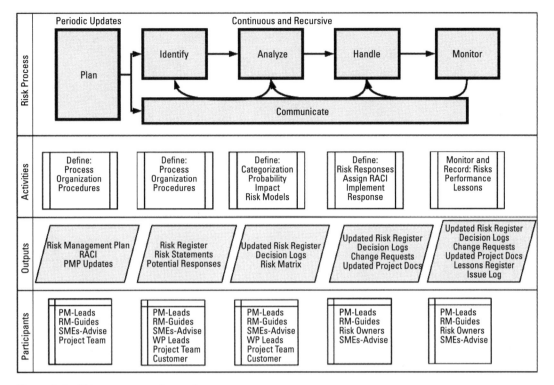

Figure 5.4 Risk management overview.

Identify Risks

The identify risk process involves finding, recognizing, and describing the risks that negatively impact the project. The objective of this process is to recognize all potential impacts to the project based on the level of knowledge currently available. Risk identification like the other PM processes is iterative because it is unrealistic to expect to know everything that could go wrong at the start of the project just as it is unrealistic to know all other aspects at that early time frame. As the project details are matured, risk identification must be revisited to validate current risks and identify new risks.

There are various identification methods to gather potential risks on projects. These include interviews, brainstorming, project document reviews, diagram techniques, and strengths, weaknesses, opportunities, and threats (SWOT) analysis, as shown in Figure 5.5 [1].

Interviews involve direct verbal communication with a range of stakeholders, senior managers, functional leads, SMEs, and team members. A good source of knowledge can be those who have previously worked on similar projects. The recommended approach is to schedule 15-minute timeslots for each one-on-one conversation and ask open-ended questions about risks. Example questions include:

- What can go wrong in this project?
- What do you see as the most challenging elements of this project?
- What areas of this project does your team have the least experience?

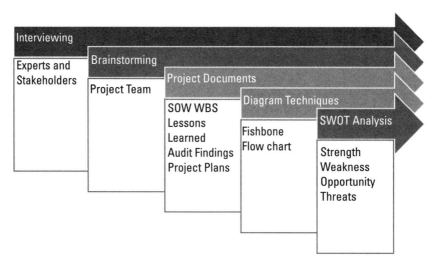

Figure 5.5 Risk identification methods.

- What potential opportunities should we consider that could help in success?

It is good to keep in mind this is a 15-minute drill, and to avoid challenging their responses, just compile everyone's stated concerns. The goal of this step is to create a list of potential risk and considerations.

Interviewing the project capture team (proposal writers) to determine if certain risks were identified upfront that can be included in the formal risk matrix is recommended.

Brainstorming involves meeting with small focus groups to identify potential risks or to expand on previously identified risks. Brainstorming requires upfront planning to address the agenda, structure, and rules of engagement for the session. Sessions involving intelligent people with strong personalities can quickly get out of control without clear structure and disciplined leadership. The goal of these sessions is to create or refine the list of potential risks and considerations. Results should be documented quickly using tools such as whiteboards, sticky notes, and easel tablets. Avoid the temptation to wordsmith or perfect the risks statements.

During the brainstorming event, the project manager should also make sure that people are respectful of the other participants and encourage building on the ideas of others. The key is to focus on quantity versus quality. It should be stated that there are no bad ideas; all should be collected then validated later. The PM should prepare for this session by generating a list of areas that should be discussed to help facilitate the discussions:

- Resource constraints (human and/or facility constraints);
- Hardware parts (long lead and/or critical parts not arriving on time, impacting schedule);
- Test and integration (failure of key performance requirements);
- Supplier risks (certificate of conformance at incoming inspection).

Project documents associated with the existing project and from other similar projects should be reviewed for potential risks. Risks may exist within these documents in the form of internal and external dependencies, assumptions, and constraints. The statement of work, project requirements, and specifications should be reviewed for cited references to government and industry standards that could represent potential project risks. Often these citations are used by the customer to cover uncertainty on their part and attempt to shift the risk to the project team.

Diagramming involves the use of fishbone diagrams, flowchart diagrams, decision trees, and models to help visualize potential cause and effects for risks.

A fishbone diagram provides the means to visualize potential issues or risks and their causes as shown in Figure 5.6. The risk is represented in the head of the diagram on the right with the potential causes decomposed to the left. Each major branch of the diagram could represent project activities, major system elements, or a combination of the people, processes, tools, and materials associated with a given project element.

A flowchart or process flow diagram maps out the general flow of processes and equipment. The network diagram created during project schedule management is an example of such a diagram. The network diagram depicts the logical flow of the project activities and the dependencies between these activities as shown in Figure 5.7. These dependencies may indicate where risks may reside in the project schedule.

The SWOT analysis method developed for process improvement can be useful here when examining and categorizing identified risks as shown in Figure 5.8. Additionally, the identification of strengths will help to identify potential opportunities. SWOT analysis is also helpful during the risk response planning described later in this chapter.

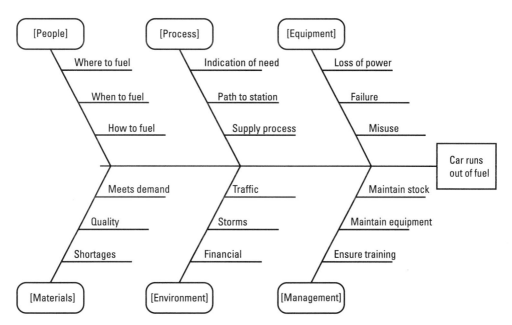

Figure 5.6 Fishbone diagram.

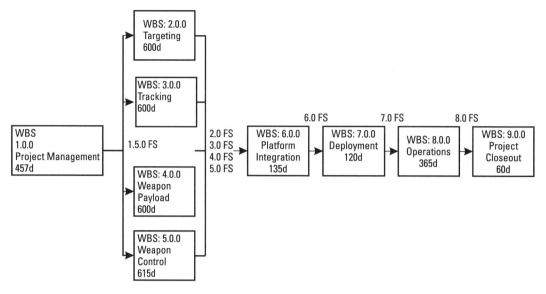

Figure 5.7   Project network diagram.

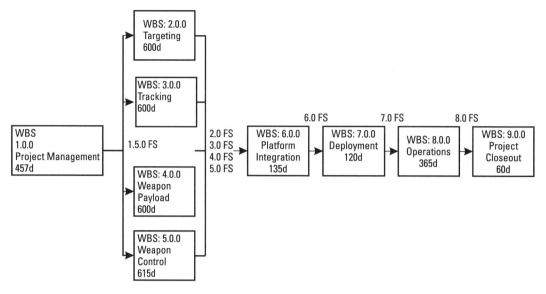

Figure 5.8   Results of SWOT analysis.

Risks cannot be managed if they are not first identified.
—Harry Hall

Using the gas station use case, an example of the identification process is as follows.

The project manager holds a risk identification kickoff meeting with her team. The meeting format is that of a brainstorming session. The PM kicks-off the meeting by explaining the goals of the meeting, the overall process and the rules for engagement for the meeting.

The PM defines the goals of the session as

- Identify the project's objectives to be used in risk identification;
- Develop a list of potential threats and opportunities that could impact these objectives.

The team starts off by discussing the project objective. One may assume the project objective is to buy gas or fill up the tank. Risk management planning focused on just filling the tank may omit the identification and prioritization of important risks. The team looks at the project charter and finds that it states the primary objective as, "establish the means to maintain an acceptable level of fuel in the car." This objective is used to begin the brainstorming for potential risks.

Understanding that fueling a car is a process used to maintain the fuel level, the team discusses the process for going to a gas station and the things that could go wrong. The following potential risks were identified:

- The gas station could be closed when the vehicle arrives;
- The station could be out of the preferred grade of gasoline;
- The station may not accept the form of payment available;
- There may be price competition between stations.

Broadening the viewpoint the team thinks about what could go wrong before arriving and added a few more risks:

- The vehicle could run out of gas before reaching the gas station;
- The road to the gas station could be closed due to an accident;
- There may be multiple fueling stations available.

Next, the team attempts to identify the cause, effect, and potential response for each risk. The results to this step depicted in Table 5.3.

This use-case illustrates the technical, regulatory, supplier, process, and human events that can impact the outcome of a project's objective. Delay's in one area (arrival, traffic) can impact other areas (run out of gas, station closed, missed appointments). Failure to plan (wrong payment, procrastination) can result in inability to execute (cannot pay, ran out of gas). The station operator running out of gas or failing to maintain his equipment is an example of risk events that the project manager can do little to prevent, yet the impact to his project still exists. In cases such as these the best response may be to minimize the impact if the risk occurs.

Risk Identification Categories

Risk categories help to identify similar risks, the resources required to respond and the ability to develop common responses to similar risks. The DoD Guidance [13] cited at the beginning of this chapter lists six common categories:

Table 5.3 Results of Preliminary Risk Definition

Risk / Event	Cause	Impact	Potential Response
Gas station closed	Unaware of hours	Cannot refuel	Be aware of hours
Gut of the preferred grade	Operator failing to meet demand	Cannot refuel with preferred	Pay higher price for better grade
Pumps out of order	Operator failing to maintain equipment	Cannot refuel	Pay higher price for higher grade
			Use alternative station
Does not accept credit card	Operator restrictions due to credit card fraud	Cannot pay with credit card	Alternative payment method
		Cannot refuel	Go somewhere else
Price competition	New discount provider across the street	Lower prices	Cost savings or ability to afford higher grade at same price
Run out of gas	Procrastination, fuel gage failure, traffic delays	Delay in proceeding to next stop, additional time and cost to have tow truck bring gas	Fuel early
			Improve fuel gauge performance
			Have alternative locations identified
Road closure	Accident, weather, construction	Run out of gas, arrive after station is closed	Have alternative routes planned, and locations identified

- Technical risks involve potential impacts due to technology readiness levels (TRL) assessments, requirements, engineering, quality, logistics, and integration. These risks can be generated internally or externally, and can impact cost, schedule, and performance.

- Technology risks are associated with research, trade studies, requirements, performance specifications, analysis, prototyping, product design, and development.

- Engineering risks are associated with engineering technical processes, technical management processes, architectural design, product and software development.

- Integration risks are associated with internal or external integrations, including customer-furnished equipment or supplier subsystems.

- Programmatic risks are potential impacts associated with cost estimates, project planning, resource constraints, project execution, communications, and project constraints.

- Business risks are potential impacts associated with resources, priorities, market forces, customers, and weather. The project manager may have business-related risks but will have little if any influence over the mitigation plan since they are external like the gas station running out of fuel. It is important to capture business-related risks and ensure management is fully aware of them. Response options could be limited when the project is impacted by certain regulations, hiring resources, or weather. Early awareness of these risks will increase the ability to make informed proactive decisions.

In the gas station example, the identified risks are grouped into three categories: technical, programmatic, and business as shown in Table 5.4.

Risk Identification Statement

After compiling risks and risk categorization, the next step is to structure risks into clear and concise statements following an "if...then..." structure as demonstrated in Figure 5.9.

Within this structure the "if" is the risk condition and the "then" is the impact of the risk occurrence. Risks also have an associated cause and effect. Using the potential risks identified in the brainstorming session the risk statements are structured as follows:

- IF requirements are not discovered and corrected by design review, THEN requirements defects will migrate into the design causing rework, impacting technical performance;
- IF the gas station does not accept the available form of payment, THEN driver will not be able to purchase gas at this station, impacting schedule;
- IF the vehicle runs out of gas before reaching the gas station, THEN a tow truck will be required to bring fuel to the vehicle, impacting schedule and cost;
- IF the road to the gas station is closed due to an accident, THEN vehicle will be delayed in arriving at the fuel source.

Risk identification is meant to informally capture all the potential problems of a project, utilizing identification methods and structuring the risks into proper "if... then..." statements as shown in Figure 5.9. This makes risk communications clear, concise, and easy to present to the team, management, and the end customers.

Table 5.4   Risk Identification Categories

Technical	Programmatic	Business (External)
Requirements	Estimates	Resources
Engineering	Program planning	Priorities
Integration	Schedule	Market
Quality	Communication	Customer
Logistics	Contract structure	Weather

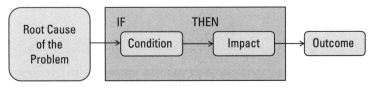

Figure 5.9   Risk identification statement (if... then...).

Analyze Risks

Analyze risks is the process that estimates the probability or likelihood of a risk occurring and the consequence to the project if the risk is realized. Risk likelihood and consequence attributes provide the structure to evaluate and assign resources required to address risks before they are realized. The likelihood and the consequence values are assigned to each risk using a graduated scale as depicted in Tables 5.5 and 5.6. Likelihood and consequence can be subjective, but the goal is to assess the risks in a consistent manner that can lead to prioritization and overall importance. The project manager works with both the risk owner and the subject matter experts to arrive at agreed upon values and scoring for each risk.

Risk likelihood defines the likelihood of a risk occurring by asking, What is the probability of the risk occurring? Likelihood levels range from the lowest likelihood (1) to the most probable (5). The key here is that all risks are scored using the consistent criteria, resulting in the ability to compare and prioritize risks.

The five levels shown in these examples can be customized based on need. Some risk managers will use a scale of 1 to 3 on small projects to simplify the process where others prefer a scale of 1 to 5.

The risk consequence defines the level of impact to the project's cost, schedule, and performance also using a graduated scale. With risk consequence, values are assigned to the potential impact on the project's cost, schedule, and performance as depicted in Table 5.6.

Risk ranking and prioritization enables risks to be presented and communicated consistently to support risk management decisions. One way of ranking risks is to map the likelihood and consequence values on a color-coded matrix like that in Figure 5.10. In this method the risk is plotted on the matrix at the intersection of its likelihood (on the y axis) and its consequence (on the y-axis). The mapping of consequence should utilize the greatest anticipated impact for the three areas. For example, the risk owner assessed the consequence for a risk with ratings of cost at level 2, schedule at level 2, and performance at level 3. In this case the highest value consequence level 3 would be used for the plot.

Using the running out of gas risk suppose the likelihood is 3 and the consequence is a 4. The risk would be mapped to the square containing the 12. This process would be repeated for each risk until the risk scores for each are identified using the matrix. The location and color of each plot denotes prioritization of the risk.

Table 5.5   Risk Likelihood Scoring Criteria

Level	Likelihood	Probability of Risk Occurring
5	Near certainty	>80% to ≤99%
4	High likely	>60% to ≤80%
3	Likely	>40% to ≤60%
2	Low likelihood	>20% to ≤40%
1	Not likely	>1% to ≤20%

Table 5.6 Consequence Scoring Criteria

Level	Cost	Schedule	Performance
5	≥10% Increase over project budget.	Cannot meet key project milestones (Slip > 'x' months)	Severe technical performance impact, cannot meet key technical threshold and will jeopardize project success.
4	5% – <10% increase over project budget.	Project critical path affected (Slip < 'x' months)	Significant technical performance impact may jeopardize project success. Work-around required to meet project tasks.
3	1% – <5% increase over project budget.	Minor schedule slip, able to meet key milestones with no schedule buffer (Slip < 'x' months)	Minor reduction in technical performance with limited impact. Work-around required to achieve project tasks.
2	<1% increase over project budget.	Able to meet key dates (Slip < 'x' months)	Reduction in technical performance, requirements can be tolerated with little or no impact.
1	Minimal cost impact. Cost expected to meet approved funding.	Minimal schedule impact	Minimal impact to technical performance.

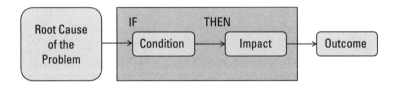

Example	
Root Cause	Vehicle operator often waits until the gas gauge is on E
If	Vehicle runs out of gas before reaching the station
Then	A tow truck will be required to bring fuel to the vehicle, impacting schedule and cost

Figure 5.10 Example of risk statement (if... then...).

High-priority risks. Risks that fall within dark shaded area in the top right of the cube with a ranking of 15 to 25 on the 5 × 5 matrix are the most critical risks that would indicate the highest need for action. During the handle risk process the project manager works with the team to develop a plan to mitigate the risks down to a lower level. The development of a contingency plan to support the impact of the risk if realized is also developed during the response planning. The project manager will need to set aside project funds to support the risk activities. It is imperative to have enough management or contingency reserve funding in case the risk becomes realized. The best way to eliminate high-priority risks is by identifying a plan to reduce the likelihood of the risk. Reducing the likelihood of the risk is usually less costly to implement than attempting to reduce or absorb the consequence.

Medium-priority risks. Risks that fall within the medium shaded area through the center of the cube with a ranking of 8 to 12 on the 5 × 5 matrix are considered moderate risks that should be addressed soon. During the handle risk process strict timelines must be established in the mitigation plan to ensure that the plan gets resolved before the risk becomes a high priority, or an issue. Note that under the 5 × 5 risk matrix, there are two color codes for risk rank number, such as:

- *Yellow*: Risk with a consequence of 5 and a likelihood of 1 will receive a medium-priority risk because the potential impact on the project is high;
- *Green*: Risk with a consequence of 1 and a likelihood of 5 will receive a low-priority risk, due to the reduced impact to the project's success.

Low-priority risks. Risks that fall within the lighter shaded areas in the bottom left of the cube with a ranking 1–6 for 5x5 matrix are low-priority risks and much less urgent than medium or high-priority risks. During the risk handling process a mitigation plan is still required for these risks but the resources applied may be delayed or reduced. Do not make the mistake of ignoring low-priority risks as unattended low-priority risks can become a high-priority issue later.

The next step is to define the risk exposure, which is the economic impact (consequence) for each risk so that resources required to respond to each risk can be identified. In other words, how much will the PM need to respond to the risk? This step requires translation of schedule and performance consequence to monetary values.

For the risk, IF the vehicle runs out of gas on the way to the station, THEN a tow truck will be required to bring fuel to the vehicle.

The cost impact could be the cost of the tow truck; about $250. On a real project it may be the cost of additional workforce, extended hours, additional materials, or rework.

The schedule impact is the lost time to wait for the tow truck, and then proceed to the gas station. In this case the schedule cost is the time the driver lost multiplied by the rate of the driver. In this case, the vehicle operator cost to the project $150 per hour and the delay was 4 hours, which equates to a consequence of $600. This is where the scheduler is needed on the risk team. The scheduler will be able to look at the critical path to determine exact impacts to the schedule.

Performance impact cost estimating can be a bit more complicated as it requires an understanding of the financial implications of not meeting a specific performance objective. Using the scoring criteria first will help to define the level of impact on the project to determine if it results in project failure, rework, or a change in the scope. The monetary value of performance consequence could range from loss of the contract, loss of incentive fees, to reduction of scope, and contract value. For this example, the impact is a loss of the $20,000 incentive award for quality standards.

In this case, the total consequence could be $20,850.

Risk exposure is based on an estimate of the probability of a risk (%) and its impact ($)

Based on the example given: 60% probability * $20,850 impact = $12,500

The project manager is now ready to present the prioritized risks to the decision-makers, as shown in Figure 5.11.

Care should be taken to ensure risks are presented with the following characteristics:

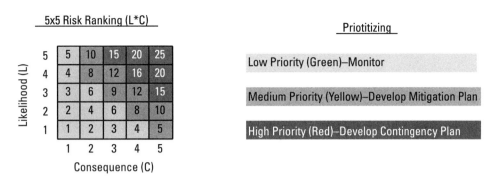

Figure 5.11   Risk ranking and prioritization.

- Risk Identified in an "if...then..." statement;
- Risk owner and the initiation of risk date assigned;
- Risk category, which impacts either cost, schedule, and/or the technical aspect of the project's success ;
- Probability or likelihood of risk occurring, and consequence anticipated;
- Risks are ranked using their plot scores and prioritized in high, medium, or low;
- Risk impact in $value is identified to cost, schedule, or performance (basis of estimate, based on actual or judgment);
- Risk exposure in $value (probability % * impact $).

Reserve funds. Project organizations typically set aside funds for supporting risk mitigation or to cover impacts caused by risk realization. These funds typically fall into one of the following categories.

- Management reserve is a portion of the project's total allocated budget set aside for management to control and not is assigned to specific tasks within the project. Management reserve is identified at the time of project cost estimating and held by management to cover the cost of unforeseen events. Often all or a portion of the reserve is held back from the PM until allocated under some sort of formal process. The risk management board illustrated in Table 5.1 is an example of this process.
- Contingency reserve is also identified at the time of project cost estimating or during project planning. Contingency reserve should be based on project risk, but are typically managed by the PM or allocated to specific control accounts within the project.

While these two terms are used throughout the project management industry, the methods used to calculate, manage, and release either of these reserves may vary from one organization to the next. For the purpose of this discussion, the term reserve funds is used as the blanket term to describe any funds help back by the PM or the PM's organization for the purpose of supporting risk management.

It is vital to identify risks during proposal development before finalizing pricing to include the cost of potential risks. As the assigned project manager might not have been part of the proposal effort, he or she must identify the level of reserve funds early in project planning.

Handle Risks

Risk handling is the decision-making process in which responses to risks are selected. Risk response options include accepting, avoiding, transferring, or mitigating each risk. The objective of this step is for the project team and project organization to establish a proactive stance in the management of risk. Depending on the complexity and visibility of the project, the authority to decide on risk responses will range from the PM to senior executives within the organization (see Table 5.7).

Before selecting a risk response these actions must be taken:

- Project risk management plan developed (Section 1.1);
- Project risk management organization established (Table 5.1);
- Risk identified and structured into appropriate statements with preliminary risk responses. (Section 5.2);
- Risk analyzed and entered in a risk register with supporting data (Section 5.3);
- Estimate a date of when risk should be mitigated and accomplished, before it becomes an issue (Section 5.3, Figure 5.12).

Risk acceptance is the response that may be chosen when all forms of response are deemed to be ineffective or infeasible. With this response, the project team acknowledges that the risk response actions are limited to those required if the risk occurs. In this case there is no proactive plan to reduce the likelihood of the risk. When a risk acceptance option is chosen, management must identify resources to support the impact if the risk is realized. In the gas station example, the road closure cannot be prevented but there can be a plan to reroute if it occurs.

Risk avoidance is the response that involves taking actions required to prevent the risk from occurring or remove any consequence if it occurs. In the gas station example, the team may elect to invest time and money into ensuring the tank never goes below half a tank or filling the tank every day to eliminate the risk of an empty tank. Risk avoidance may be the most expensive of all risk response options.

Risk transfer is shifting the burden of a risk to another entity. Transferring a risk to another organization may result in loss of control and will most likely come at a cost. For example, the project manager may elect to shift the responsibility of filling the tank to a third party that delivers fuel to the vehicles so that the tank is always full. Be cautious with this approach as the project manager has ownership

Table 5.7   Handle Risk Strategies

Accept	Avoid	Transfer	Mitigate
Acknowledge that the risk event or condition may be realized.	Reduce or eliminate the risk event or condition by taking an alternate path.	Reassign the risk responsibility to an external party.	Seek to actively reduce the likelihood and/or consequence of the risk to an acceptable level.

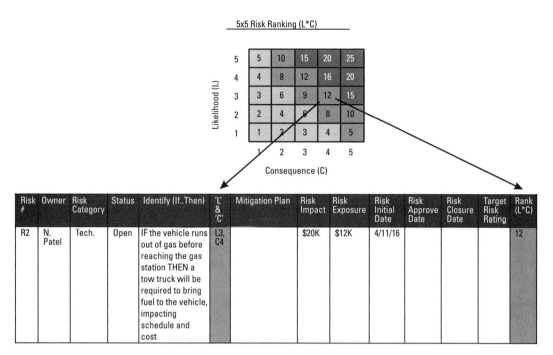

Figure 5.12   Risk register.

and responsibility for managing suppliers and the team to deliver successful products. Not all risks can be transferred. Inclement weather, political unrest, outsourcing, and labor strikes are some examples of events that can significantly impact the project's success and are out of the control of the project team and not possible to transfer to another entity.

Risk mitigation is the response that attempts to control risk impact. Risk mitigation requires a plan of action that will drive the risk likelihood down as low as possible and to minimize the consequence of risk realization. This response requires a mitigation plan that details the activities required to reduce the likelihood as low as possible and minimize potential project impacts (see Figure 5.13). Keep in mind that the best and least costly way to reduce risk is to reduce likelihood. Having a good plan in place and creating risk awareness is the key to enabling mitigation.

Identifying risks as early as possible in the project supports accurate project estimates and the creation of reserve funds. The plan of action associated with each risk response should include

- Contingency plans for responding to risk occurrence;
- Allocation of budget and schedule reserves to address consequence;
- Key resources (SM's) identified;
- Monitoring for risk indicators before occurrence;
- Overcommunication of risk with key stakeholders to increase awareness and proactive attitudes;
- Process improvements for risks due to process concerns;
- Create standards to guide company practices;

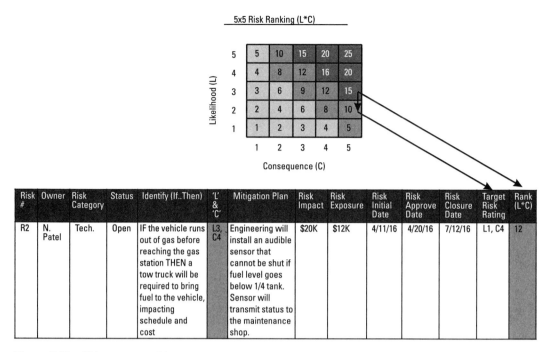

Figure 5.13   Risk response-mitigate.

- Redundancy to eliminate any single-point failure;
- Training to increase work quality.

Once a decision is made to mitigate the risk, the project manager should develop a detailed burn-down plan of all high and medium risks. Tracking progress of the risk can be performed as shown in Figure 5.13.

Create a measurable action plan to show how risk can be reduced to an acceptable level or closure over time. Ensure each activity is clear, specific, and measurable to aid in reporting and tracking. In Figure 5.13, the bottom left quad provides an example of detailed risk mitigation action steps. It illustrates how each step reduces the likelihood of a risk and identifies each action owner, as well as due dates for each action. The bottom right quad of the figure illustrates how risk becomes burned-down over a specified timeline. In the top right matrixes displays risk in terms of likelihood and consequence to demonstrate how risk can be reduced.

The objective of risk handling is to determine which action should be taken to reduce the likelihood or consequence of risk by either accepting, avoiding, transferring, or mitigating the risk. Most often the risk mitigation option is selected to reduce risk to an acceptable level. Risk avoidance and transfer are the lesser used option, due to costs associated with shifting the risk to other organizations.

Monitor Risks

Risk monitoring is the process of recording, maintaining, and reporting risk status along with the continuous analysis, handling, and tracking of risk. Risks should always be presented during program reviews, technical reviews, working groups, or risk management boards. Risk monitoring tracks all identified, residual, and new risks. The process continues throughout the entire project life cycle. New risks are identified and current risks come to closure. It is important to track risks periodically to ensure they are not being ignored and escalating into a real problem. This approach allows identified risks to be reduced to an acceptable level, and new risks are continuously captured, allowing the team to make proactive decisions.

Reassess all risks as required and ensure full management support is available to support risk response. This activity is continuous throughout the project and requires considerable amounts of time and resources to ensure risks are properly monitored.

- Reassess risk identification, root cause, assumptions, and modify them as necessary;
- Review and continuously reevaluate the risk mitigation plan, probability of the risk occurring, and the impact to project success;
- Encourage awareness of high and medium risks to promote continuous improvement and to obtain leadership support if risk requires urgent attention;
- Involve all levels of management in the process;
- Ensure risk owner takes responsibility and accountability for managing risks;
- Review retired risks to ensure such risks have not relapsed.

You have to know the past to understand the present.
—Carl Sagan.

Risk monitoring is the last major step in the risk management process, but it does not end until the project ends. It ensures risks are continually identified, analyzed, handled, and monitored. By properly implementing risk handling mitigation plans, risks can be reduced to an acceptable project level.

Opportunity Management

Project opportunities are the opposite of project risks. Things may occur that will benefit the project's likelihood of success. To identify and analyze such opportunities, follow the same process as identifying and analyzing a risk. The brainstorming session for the gas station included the identification of several opportunities. If the opportunity occurs, then it can have a positive impact to cost, schedule, or performance. One of the key reasons to identify and track opportunities is to capture activities that could offset cost or schedule impacts due to realized risks.

Once the opportunity is identified in an "if... then..." statement, the next step is to handle that opportunity as described earlier in this section. Table 5.8 translates the response strategies for threats into strategies for opportunities.

Ignore means to do nothing, making no changes to affect either probability or impact of the opportunity. When the opportunity does not present enough value, the best approach may be to monitor until conditions change.

Risk# (R1)	
Root Cause	Vehicle operator often waits until the gas gauge is on E
If	Vehicle runs out of gas before reaching the station
Then	A tow truck will be required to bring fuel to the vehicle, impacting scheduling and cost

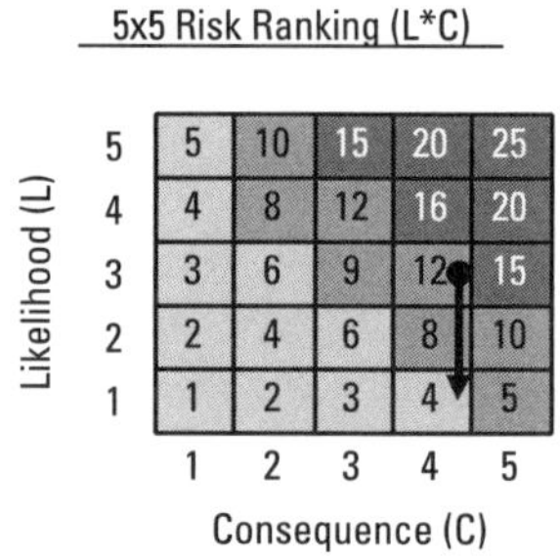

5x5 Risk Ranking (L*C)

Likelihood (L)					
5	5	10	15	20	25
4	4	8	12	16	20
3	3	6	9	12	15
2	2	4	6	8	10
1	1	2	3	4	5
	1	2	3	4	5

Consequence (C)

Step#	Mitigation Action	Due Date	Owner	L , C
1	Finalize requirements verification and validation metrics	5/12/16	E. Roulo	L3, C4
2	Conduct internal requirements review with all key stakeholders	6/10/16	E. Roulo	L2, C4
3	Conduct final requirements flow down meeting to communicate approved requirement gaps closure with each key stakeholder	6/28/16	N. Patel	L1, C4 L1, C4

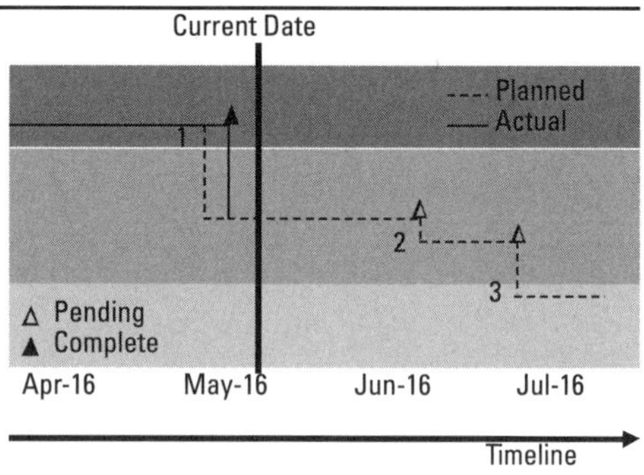

Current Date

---- Planned
—— Actual

△ Pending
▲ Complete

Apr-16   May-16   Jun-16   Jul-16

Timeline

Figure 5.14   Risk burn-down.

Exploit means to make sure every effort is made to recognize an opportunity and make it occur. Examples could include procuring commercial of the shelf (COTS) items to reduce the cost of developing a product internally or using existing technology to reduce the cost of development.

Opportunity share means to team or partner with other organizations to share cost, resources, and technology to increase an opportunity's probability of occurrence.

Opportunity enhance means to increase the likelihood of an opportunity to occur. This could be accomplished by adding skilled and efficient resources to reduce time or train people to improve quality.

Summary

Managing risk effectively is vital for any project's success. Risk management is the effort of identifying, analyzing, handling, and monitoring risks to allow the project team to be proactive, to uncertainty instead of reactive. A risk management plan defines how to handle and assess risks defines, the roles and responsibilities of team members, and the frequency of risk reporting.

Identify risks involves finding, recognizing, and describing the risks that could affect the achievement of a project's objectives. When initially identifying risks, the goal is to capture all possible concerns for the project before using the filtering process. During interviews ask open-ended questions that will help to identify potential treats and opportunities for the project.

Analyze risks measures the probability or likelihood of a risk occurring and its potential impact. This step assists with analyzing and prioritizing risks.

Table 5.8   Opportunity Responses

Threats	*Opportunities*
Accept	Ignore
Avoid	Exploit
Transfer	Share
Mitigate	Enhance

Handle risks is the decision-making process to either avoid, accept, transfer, or mitigate risk (Table 5.8). Failure to develop and implement risk response strategies results in the reduced likelihood of the project achieving its objectives.

Monitor risks means to improve the program by managing and reporting risk, along with continually analyzing risks, handling, and tracking risk results.

The following definitions for risk are drawn from the resources cited at the beginning of the chapter.

PMI: Project risk is an uncertain event or condition which, if it occurs, has a positive or negative impact on one or more project objectives such as scope, schedule, cost, and quality. [2]

DoD: Risk is a measure of future uncertainties in achieving program performance goals and objectives within defined cost, schedule and performance constraints. *Risk Management Guide for DoD Acquisition* [1].

NASA: "In order to foster proactive risk management, NPR 8000.4A integrates two complementary processes, Risk-Informed Decision Making (RIDM) and Continuous Risk Management (CRM), into a single coherent framework. The RIDM process addresses the risk-informed selection of decision alternatives to assure effective approaches to achieving objectives, and the CRM process addresses implementation of the selected alternative to assure that requirements are met. These two aspects work together to assure effective risk management as NASA programs and projects are conceived, developed, and executed" [4].

References

[1]   U.S. Deparment of Defense, Undersecretary of Acquisition, *Risk Management Guide for DoD Acquisitions,* Sixth Edition (Version 1.0), Washington, DC: U.S. Desprtment of Defense, 2006.

[2]   Project Management Institute, Inc., *A Guide to the Project Management Body of Knowledge: PMBoK Guide*, Sixth Edition, Newtown Square, PA: Project Management Institute, Inc., 2017.

[3]   Lorenz, R., and D. Harland, *Space Systems Failure. Disasters and Rescues of Satellites, Rockets and Space Probes*, Chichester, UK: Springer, 2005.

[4]   NASA, *NASA Risk Management Handbook*, 2006.

CHAPTER 6

Project Resource Management

Effective project managers require a firm understanding of the processes and activities applied to plan, execute, monitor, and control the work required. An effective PM also requires a set of soft skills to effectively lead the project team in the execution of their duties. Resource management is the set of activities performed by the PM to acquire, organize, and enable the human and physical resources associated with the project. While resource management does involve specific activities, this area of project management requires a strong understanding of the soft skills used to recruit, motivate, and retain the best available talent for the project.

The people that make up the project team can come from organizations both internal and external to the PM's organization. While vendors are the prominent type of external organization, other external organizations may provide resources to the project. Examples of these organizations include government agencies, nonprofits, other corporate divisions, and corporate partnerships. In these latter cases the people and resources that support a project may do so with or without the exchange of funds and the PM's ability to control the performance of these external resources is likely limited.

This chapter provides guidance on the use of the soft skills to plan for, acquire, develop, and manage teams of human resources and physical resources used by the team to execute the project. The objective of resource management is to have the required resources available to the project when and where they are needed. Project managers are encouraged to be familiar with the *PMBOK Guide* [1], which was used as a resource for this chapter.

Figure 6.1 depicts the top-level activities performed to plan, acquire, develop, and manage resources assigned to a project.

Figure 6.1 conveys the message that resource management is one of a set of integrated project management processes described within this publication that are used by effective project managers. Each of these processes in Figure 6.1 and their outcomes are dependent on the other processes in this figure. Effective resource management is dependent on integration with all PM processes.

Organizational Environment

Before beginning a discussion on resource management, it is important to first discuss the roles and responsibilities associated with the assignment of resources to

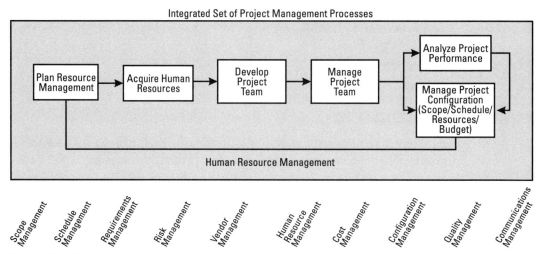

Figure 6.1   Resource management processes.

projects. Each resource within an organization typically has a single person accountable for assigning resources to projects. For human resources, this entity may or may not be the supervisor of the person or persons needed for the project. The responsibility of recruiting, training, and equipping these human resources may fall with a different party than the party accountable for assigning people to projects.

There are also parties accountable for assigning buildings, labs, and production lines to projects. The parties responsible for maintaining these facilities may consist of people separate from those who assign and use the facilities. Where the accountability and responsibility for each of these roles reside is dependent on the structure and policies of the organization in which the resources reside.

> *Responsible: The parties who do the work required.*
> *Accountable: The party charged to ensure the work is completed.*

The following describes the roles and responsibilities associated with resource management. The names applied to these roles and responsibilities may not match those used within the reader's organization. But the characteristics of these roles and responsibilities do exist and do impact the PM's ability to obtain and utilize resources. Therefore, new technical PMs can reduce their stress and frustration by understanding the environment in which their project resides and whom within the environment has the responsibility and accountability to provide resources to project.

Administrative control: The responsibility for hiring, developing, and sustaining human talent. The entity with this type of control is charged with developing forecasts for resources and then hiring or training people to support the forecasts. For example, engineering department, vendor X, and agency Y (also known as competency, group, or branch).

Functional manager: A person within the organization that is accountable for the assignment of human resources to projects. The functional manager hires, trains, and equips human resources to support business operations or projects. Typically, the functional manager is the supervisor of the human resource and must approve assignments to a project. For example, supervisor within the engineering department (also known as resource manager, branch head, supervisor, competency manager).

Operational control: The organizational unit that is responsible for the day-to-day execution of one or more business outcomes. The entity with operational control typically has responsibility and authority over facilities, utilities, safety, and support activities. Operational control can be assigned by line of business, facility type, or geographic location. A key facet of this role is the focus on the successful execution the organization's business or mission unlike a project that is a temporary endeavor to deliver a solution to a business or mission need. For example, the widget division, the widget factory, and the widget laboratory.

Operations manager: The person within the organization that has accountability and responsibility over day-to-day operations of facilities. The assignment of facility resources typically is approved by the operations manager. For example, factory manager, laboratory manager, base commander, and VP of operations.

Project control: The entity that is accountable for the planning, execution, management, and completion of one or more projects. The person with project control has the ability to utilize resources for the completion of one or more project objectives.

Organization types: Common organizational constructs are numerous, and each organization has its own variant. Additionally, reorganizations are a common occurrence within any entity. Coupled with the constantly changing leadership trends that change the organizational methods in vogue at any given time, it would be a waste of the reader's time to dive into every possible structure. Therefore, the following types are presented as the top-level characteristics found within organizations. These characteristics will help the new technical PM to identify and understand how his or her organization is structured.

> *PMBOK Guide, Chapter 2 provides valuable information on considering the environment in which a project operates and the factors these environments create that can impact a project [1].*

> *Additional information on how to manage projects associated with organization change can be found in Managing Change in an Organization. The Project Management Institute (PMI) [2].*

Functional: An organizational structure arranged according to the functions performed within the organization (see Figure 6.2). Examples of functions are research, engineering, development, production, and finance. Each functional group has administrative control of its human resources. Projects executed within a functional organization may move through each organizational unit during the project life cycle. In these cases, the functional manager for each unit may assume project control for the work executed within their functional area. The project manager on this type of project may have little or no project control and focuses on integrating

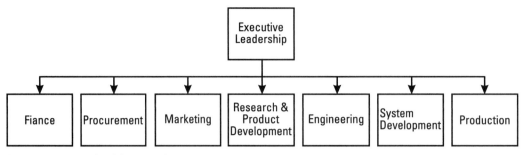

Figure 6.2 Example of functional organization.

the plans created by each functional manager. During project execution, the PM relies on progress data provided by each functional manager to track progress. The human resources within functional organizations tend to specialize in each area leading to a high level of expertise.

Projectized: An organizational structure in which resources are grouped according to one or more projects (see Figure 6.3). In projectized organizations, administrative and project control are assigned to the organizational entity executing the project. The project organization is composed of all required human resources

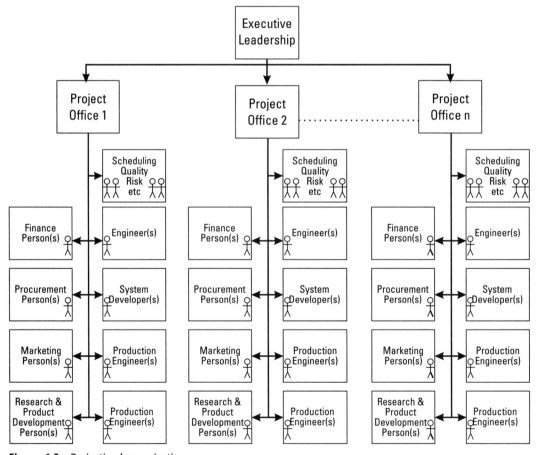

Figure 6.3 Projectized organization.

and skill sets. In this model, human resources may be expected to work across multiple disciplines resulting in a team of generalists instead of experts. The PM typically has administrative and project control in these cases. Functional managers residing within a projectized organization are typically limited to the administrative functions associated with hiring, training, and caring for the human resources.

Matrix organizations: The matrix organization is a composite of the functional and projectized types as shown in Figure 6.4. As the name implies, projects and functions are organizational entities that coexist within the same organization. Projects require a broad set of human resources, each with specialized knowledge and skills that are provided to the project by functional groups. The project manager leads an integrated project or product team (IPT) composed of only the resources required to execute the tasks within the project. The functional manager has a pool of expertise developed according to a common set of processes and tools used by her resources. Resources are assigned to one or more projects within the projectized group according to the demands for each project. During downtime the resource can move to other tasks within the same projectized group, reducing the financial burden on the first project. In the matrix organization the PM has project control and the functional manager has administrative control over the people assigned to the IPT. Additionally, the functional manager will have influence over how work is performed by his or her employees by defining standard processes, tools, and techniques for her functional area.

Matrix organizations are sometimes referred to as balanced, strong, and weak. These terms are used to describe the balance of control between the PM and the functional managers. In a balanced matrix, both parties must share in the decisions regarding resources. In a strong matrix, the PM has final say on all resources and

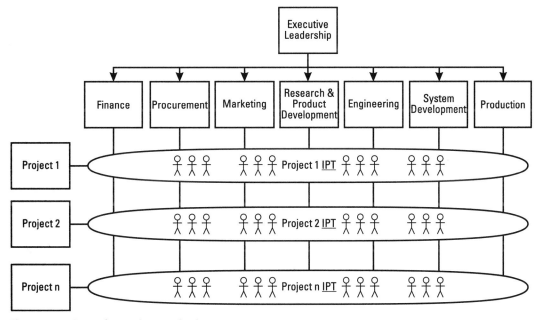

Figure 6.4 Example matrix organization.

in the weak matrix the functional managers can make resource decisions without approval of the PM.

The roles depicted in Table 6.1 demonstrate the integration between project manager and the roles that provide resources to the project. It is important for the PM to understand the accountability and responsibilities for each resource management role within the project team.

Plan Resource Management

When planning resource management, the PM establishes a strategy for how resources will be acquired based on his understanding of the work and the resources available to the project. The goal of this step is to document all resources required and to establish a strategy to acquire, develop, and manage each resource. The resource acquisition strategy defines the methods that will be used to acquire human and physical resources for the project.

During project scheduling, the PM works with his SMEs to define the tasks to be completed and the types of resources required to complete these tasks. The PM

Table 6.1 Human Resource Management Roles

Role	Responsibility
Customer lead	The member of customer's organization charged to manage the project from the customer's perspective. Has final authority over all additions, changes, or deletions to project scope, schedule, and performance.
Project manager	The person accountable for project success, the PM leads, monitors, and supports the project's resource management processes and procedures.
Procurement officer	Member of the project manager's team that can authorize new agreements and changes to agreements between the project team and the customer or between the project team and its vendors. The procurement officer also ensures all changes that impact the current agreements are executed according to the organization's policies.
Resource provider	The organization or entity providing resources to the project. Resource providers can be internal or external to the organization.
Functional manager	A person within the organization that has administrative control over a set of human resources. The functional manager hires, trains, and equips human resources to support business operations and/or projects. Typically, the functional manager is the supervisor of the human resource and must approve assignments, travel, training, and time off. Example: Supervisor within the engineering department. (Also known as resource manager, branch head, supervisor, competency manager.)
Operations manager	The person within the organization that has control and responsibility over day-to-day operations. For this discussion the operations manager has control over the use, safety, and assignment of factories, laboratories, or specialized equipment.
Vendor functional manager	A person within the vendor organization that has administrative control over a set of human resources assigned to a contract. The functional manager hires, trains, and equips human resources to support business operations or projects. Typically, the functional manager is the supervisor of the human resource and must approve assignments, travel, training, and time off.
Work package lead	The person assigned responsibility and accountability to deliver a work package. May be a member of the PM's organization or the vendor organization if the work package is outsourced.
Project scheduler	Member of the project manager's team that performs the consolidation, structuring, publication, and management of the project schedule. The project scheduler updates the schedule to reflect the amount of resources assigned to each project task.

also works with his project scheduler to determine the time frames in which each type of resource is required. The resource requirements generated during schedule development are used as the basis for the resource acquisition strategy.

The PM first determines whether the resources required are available internal to his organization or if an external source must be identified. During the risk identification process, factors such as level of expertise, capacity, and task complexity are identified that serve to help in this process. For example, a portion of the project involves developing a software package in which the organization has limited experience. A risk response plan may be to hire an outside vendor with demonstrated experience for this work. In another case, the project requires specialized equipment or production facilities to support the project. The PM and his organization may decide to outsource this segment of work rather than build their own production capability. There will also be work that is best or can only be performed by internal resources. Financial management, marketing, procurement, and systems engineering are examples of this type of work.

Resource needs deemed to be sourced from external sources will require the procurement process and authorities defined in Chapter 7. The acquisition strategy will reflect the need to use the procurement officer to lead the vendor acquisition processes.

Resourced needs deemed to be sourced internally will follow a different set of processes described later in Section 6.3.

The arrangements made to commit resources to a project can follow one of the following:

- *Dedicated:* Resources assigned to the projected are committed to the project for all or a portion of the work week for an agreed time frame. The people assigned to the project are expected to be available to the PM and working on the project for the agreed amount of time each week. Dedicating a resource to a project does not mean a transfer of administrative control to the PM.

- *Work package delegation:* Work is delegated to individual departments, contractors, or managers to be completed. The work package is completed using a mix of resources within the work package owner's organization. In this method, the work package owner provides an estimate for the time and resources required to support the entire work package. The work package owner may use his resources across multiple tasks and can surge resources in and out of the project as needed.

- *Support-as-a-service:* This method is often used for internal transactions between the project and indirect support functions such as procurement, information technology (IT), and facilities. In this fee for service structure, the party providing support to the project charges a fee for their services, which are executed by a mix of resources within the service provider's organization. This method of acquiring IT capabilities using a fee-for-service model is growing in prominence due to the expense of acquiring and maintaining complex IT infrastructure. In this model, the infrastructure cost is shifted to the service provider, freeing capital investment funds for items specific to the organization's business or mission.

These arrangement types should not be confused with the types of contracts outlined in Chapter 7. The contract types are the structure by which costs are charged to the project under the formal agreement between the project and the vendor. The work packages and staffing arrangements described above could be contracted using firm-fixed price, cost plus, or time and material contracts. Additionally, transactions between internal operating units can also have varying methods for calculating and charging costs to projects.

There are positive and negative aspects associated with these approaches and the applicability of these aspects will change from project to project. Every project is unique and what works for one project may not work for a seemingly similar project. A few of factors to consider when deciding on the resourcing approach are as follows.

PM workload: Does the PM have the time and bandwidth required to recruit, screen, and directly manage every member of the project team? In most cases the PM has multiple small projects running concurrently or a single large-scale project with a multitude of resources. In either case, the odds that the PM has the time required to directly manage resources and the work performed by resources are rare in today's business environment.

> *An effective PM learns how to delegate and monitor work instead of managing worker activities.*

Level of commitment: There may be cases where the best way to ensure work is completed on time and within budget is to dedicate resources to the project. The resources are assigned to the project and the project pays for all the resource's time regardless of their workload. At the same time, the person is dedicated to the project and the PM does not need to worry about the resource being available when needed. Dedicating resources is often required during the initial planning stages of the project.

Surge capacity: There will be cases where the amount of work required exceeds the time allotted for the persons assigned to a task. The reality of project management is that something will go wrong or take longer than scheduled, impacting the start date of downstream tasks. Assuming the project completion date is inflexible, the delays must be made up for in order to meet the project's completion date. Surge capacity is the ability to add new resources to the project during peak periods. The potential need for surge capacity should be discussed with each resource provider to determine the best method for staffing the team.

Fragmented tasks: There will be activities in the project that have a surge of activity followed by a waiting period until the resource is required for the next task. Unless the project has numerous similar tasks, the resource may have wait times, which equates to unproductivity.

Resource workload: A resource's workload is based on the total work assigned and the time frame for when the assignments must be completed. Resource workload must be managed throughout the project. A resource may appear to have a balanced workload until a predecessor or current task slips causing the resource to have multiple concurrent tasks. A resource that starts off fully loaded with no surge capacity will not be able to absorb slippages in predecessor tasks.

Clarity of scope/work packages: Despite best efforts, there will be cases where the work is not defined sufficiently to issue work packages for portions of the work. This will be the case particularly during project initiation. In these cases, resources may have to be dedicated to the team until work can be planned into assignable work packages.

Organizational constructs: The organizational constructs discussed earlier will have a heavy influence on how projects are resourced. Some of these decisions may be outside of the PM's control but the project must be planned, documented, and monitored just the same by the PM.

There is no perfect answer or best strategy for identifying and recruiting resources. In fact, each project may have a combination of work packages, dedicated resources, and support-as-a-service arrangements. The PM improves the likelihood of making sound decisions by understanding the project scope, identifying the unique characteristics of each work element, and communicating this information to the resource providers.

For each work element or group of work elements, the PM works with his SMEs to identify the resource requirements and the strategy best suited to acquire the resources. Each task is assigned a resource requirement, a designation for internal or external sourcing, the arrangement planned for the resource, and the projected provider. Included in this data set is the time frame in which each task is scheduled to be completed. Figure 6,5 demonstrates the first round of this process where a strategy was documented for several elements of work based on the schedule developed in Chapter 3.

Upon completing the resource management planning process, the PM now has a clear understanding of the types of resources required for each element of work within the project plan. The PM also has an initial strategy for how he will acquire the resources and the method to be used to commit resources to the project.

Acquire Project Resources

The goal of this process is to obtain agreements from the resource providers to commit human and physical resources to the project. Depending on the acquisition method identified during the planning stage, one of three principle paths will be followed to acquire the resources.

1. Procurement Initiation according to the vendor management processes described in Chapter 7;
2. A human resource hiring action according to the organization's hiring processes;
3. Resource negotiation and assignment with internal functional managers.

The need for clear and concise requirements applies regardless of the organizational construct or the acquisition method planned. Procurement packages, recruitment packages, and internal work assignments all require clear and concise requirements to be effectively satisfied.

The PM uses the data created during the acquisition strategy described in the previous step to communicate his needs to the resource providers. However, the data as presented in Figure 6.5 is not structured to communicate these needs to each provider. In Figure 6.6 acquisition strategy data is now sorted by resource type, which helps each resource provider quickly identify their area of expertise.

Additional views of this data can be created to provide a list of all contract actions, internal resource needs, or other views customized to the audience. Providing clear and quickly identifiable resource needs to the providers will improve their ability to respond.

WBS	Task Name	Start (Planned)	Finish (Planned)	Resource Requirement	Internal / External	Method	Provider
0.0.0	Q36 Space Modulator	07/02/18	12/15/25				
1.0.0	Project Management	07/02/18	12/15/25				
1.1.0	Project Initiation	07/02/18	07/02/18	Program Manager	Internal	Dedicated	Program Office
1.2.0	Project Charter	07/02/18	08/31/18	Project Manager	Internal	Dedicated	Project Management Dep
1.3.0	Schedule Management	09/03/18	05/10/19	Scheduler	External	Dedicated Contractor	TBD
1.4.0	Planning Documents	02/18/19	06/21/19	Project Manager	Internal	Dedicated	Project Management Dep
1.5.0	Work Authorizations	04/01/19	09/13/19	Procurement Official	Internal	Support as a Service	Project Management Dep
1.6.0	Monitoring and Control	04/01/19	12/29/25	Project Manager	Internal	Dedicated	Project Management Dep
2.0.0	Targeting	06/24/19	10/22/21				
2.1.0	Concept Development	06/24/19	08/14/20				
2.1.1	Functional Requirements	06/24/19	09/13/19	Requirements Manager	Internal	Dedicated	Systems Engineering
2.1.2	Architecture	09/16/19	12/06/19	Systems Engineer	Internal	Internal Work Package	Systems Engineering
2.1.3	Systems Analysis	12/09/19	02/28/20	Systems Analyst	Internal	Internal Work Package	Systems Engineering
2.1.4	Solutions Analysis	03/02/20	05/22/20	Systems Analyst	Internal	Dedicated	Systems Engineering
2.1.5	Prototyping	05/25/20	08/14/20	Turnkey System Prototyping	External	Contract Work Package	TBD
2.2.0	Design and Development	08/17/20	07/16/21				
2.2.1	System Requirements	08/17/20	11/06/20	Requirements Manager	Internal	Dedicated	Systems Engineering
2.2.2	System Design	11/09/20	01/29/21	Systems Engineer	Internal	Internal Work Package	Systems Engineering
2.2.3	Specifications	11/09/20	01/29/21	Systems Engineer	Internal	Internal Work Package	Systems Engineering
2.2.4	Sub-systems	02/01/21	04/23/21	Turnkey Design	External	Contract Work Package	TBD
2.2.5	Software	02/01/21	04/23/21	SW Engineers	External	Contract Work Package	TBD
2.2.6	Integration	04/26/21	07/16/21	Integration Shop	External	Contract Work Package	TBD
2.3.0	Test and Certification	04/26/21	10/22/21				
2.3.1	Developmental Tests	04/26/21	08/06/21	Test Director	Internal	Dedicated	T&E Group
4.3.2	Technical Readiness	07/19/21	09/03/21	Test Director	Internal	Dedicated	T&E Group
4.3.3	Operational Readiness	09/06/21	10/22/21	Test Director	Internal	Dedicated	T&E Group

Figure 6.5   Example of resource acquisition strategy by work.

Resource Requirement	Provider	Internal / External	Method	WBS	Task	Start (Planned)	Finish (Planned)
Integration Shop	TBD	External	Contract Work Package	2.2.6	Integration	04/26/21	07/16/21
Procurement Official	Project Management Dep	Internal	Support as a Service	1.5.0	Work Authorizations	04/01/19	09/13/19
Program Manager	Program Office	Internal	Dedicated	1.1.0	Project Initiation	07/02/18	07/02/18
Project Manager	Project Management Dep	Internal	Dedicated	1.2.0	Project Charter	07/02/18	08/31/18
Project Manager	Project Management Dep	Internal	Dedicated	1.4.0	Planning Documents	02/18/19	06/21/19
Project Manager	Project Management Dep	Internal	Dedicated	1.6.0	Monitoring and Control	04/01/19	12/29/25
Requirements Manager	Systems Engineering	Internal	Dedicated	2.1.1	Functional Requirements	06/24/19	09/13/19
Requirements Manager	Systems Engineering	Internal	Dedicated	2.2.1	System Requirements	08/17/20	11/06/20
Scheduler	TBD	External	Dedicated Contractor	1.3.0	Schedule Management	09/03/18	05/10/19
SW Engineers	TBD	External	Contract Work Package	2.2.5	Software	02/01/21	04/23/21
Systems Analyst	Systems Engineering	Internal	Internal Work Package	2.1.3	Systems Analysis	12/09/19	02/28/20
Systems Analyst	Systems Engineering	Internal	Dedicated	2.1.4	Solutions Analysis	03/02/20	05/22/20
Systems Engineer	Systems Engineering	Internal	Internal Work Package	2.1.2	Architecture	09/16/19	12/06/19
Systems Engineer	Systems Engineering	Internal	Internal Work Package	2.2.2	System Design	11/09/20	01/29/21
Systems Engineer	Systems Engineering	Internal	Internal Work Package	2.2.3	Specifications	11/09/20	01/29/21
Test Director	T&E Group	Internal	Dedicated	2.3.1	Developmental Tests	04/26/21	08/06/21
Test Director	T&E Group	Internal	Dedicated	4.3.2	Technical Readiness	07/19/21	09/03/21
Test Director	T&E Group	Internal	Dedicated	4.3.3	Operational Readiness	09/06/21	10/22/21
Turnkey Design	TBD	External	Contract Work Package	2.2.4	Sub-systems	02/01/21	04/23/21
Turnkey System Prototyping	TBD	External	Contract Work Package	2.1.5	Prototyping	05/25/20	08/14/20

Figure 6.6   Example of work rollup by resource type.

Internal resources: It may come as a surprise to new project managers that the assignment of internal resources to projects can often be the most stressful. Functional managers are frequently challenged with recruiting, developing, and retaining the level of qualified talent that matches the current level of demand.

Developing an inexperienced employee into a valued project team member requires years of training, coaching, and experience. Recruiting experienced employees requires convincing people that changing employers to join the manager's team is worth the change. These experienced people come at premium rates that must be supported by a steady flow of quality work. The quality of work, the working conditions, and a balance between work and home life are critical factors for employee retention. For this reason, a functional manager may place a priority on the employee's job satisfaction and allow an unhappy employee to change assignments to avoid losing the employee.

Recruiting, hiring, and onboarding new talent for an organization can be a lengthy process. At the same time, new projects often come into an organization with little advanced notice to the resource providers. Also the same time, new projects are expected to start and show results in short order. For these reasons, PMs often find themselves competing with their peers to obtain the best talent for their projects from a limited supply of resources. The following techniques can aid the PM in obtaining and retaining qualified talent.

Maintain strong relationships: Relationships are important to project success. Maintaining open lines of communication can help to build strong relationships with resource providers. This open line of communications is critical when asking for additional resources or addressing performance issues. One should not underestimate the power of positive feedback to managers. A PM often knows more about the day-to-day performance of an employee than their supervisor. Positive feedback provides opportunities to strengthen relationships with the employee and the resource manager.

Build partnerships: Establishing mutually beneficial relationships with the functional managers within the resource provider organizations can be a motivator when supporting projects. Functional managers are typically experts in their discipline and are motivated by opportunities to develop the people, processes, and tools within their discipline. A PM can build strong partnerships by understanding the motivations of his or her functional managers and seeking opportunities to support their goals. Leaning on the expertise of these subject matter experts throughout the project life cycle is another way to demonstrate respect for their talents and further strengthen the partnership.

Understand roles: Respecting the roles and responsibilities between the PM, the assigned resource, and the functional manager will avoid strains on these relationships. An employee that feels unsupported or receives conflicting directions will look for other opportunities and frustrated managers will be reluctant to provide resources to future projects.

Communicate changes early: The earlier a resource provider is aware of a potential change in requirements the faster she can respond. Communicating potential new projects or changes to existing projects to internal providers as soon as possible enables the provider to work on long lead time resources. Communicating potential project delays or work stoppages is equally important if the resource provider is expected to cover the employee during the delay.

Focus on requirements: The PM is the customer of the resource provider and the resource provider, like the PM prefers to deliver solutions to requirements not be told what solution to provide. Providing clear and concise requirements helps the resource provider understand the project's needs. Allowing the resource provider to match her resources to the project needs, rather than specific demands, gives the resource provider flexibility in meeting the PM's needs.

Grow talent from within: Every PM wants the best and brightest talent assigned to his projects. The reality is that the best and brightest come in short supply. When trying to resolve resource shortfalls, an assumed solution is to hire additional people or outsource the task. Establishing a new contract or hiring a new employee can be time-consuming in any organization. Allowing the resource provider to develop an existing member of her team into the type of talent required is an example of the partnering technique. This approach helps the functional manager build her talent base and strengthens the partnership.

Be flexible: Many people have talents beyond their current position. Often the talents required to do one type of work is like that of another discipline. For example, test engineers often have experience with requirements analysis and user interface design. In other cases, a potential resource may aspire to learn a new talent. Being flexible in the approach to acquiring resources helps to build relationships while achieving required resources.

The process to acquire internal resources should be defined within the PM's organizational processes assets. Many organizations have processes and tools to communicate resource requirements to functional managers as well as for communicating future demand. The resource acquisition strategies developed during the planning process and the techniques described above can lead to efficient and effective team building in accordance with the organization's processes.

The last step in resource acquisition is to obtain commitments from each provider. These commitments should reflect the tasks, time frames, and any tools or facilities required by each resource. Depending on the lead time, it may not be realistic for the agreement to the name a specific resource. In these cases, it is imperative that the agreement reflect the knowledge, skills, and abilities (qualifications) expected under the commitment. Additionally, any actions such as recruitment or capital investments must also be documented in the agreement and incorporated into the project management plan.

Develop Project Team

Human resources are assigned to the team from numerous sources, geographic locations, and cultural backgrounds. Additionally, each person has unique personality and generational characteristics and are likely in different stages of career development. The PM must now mold this collective of individuals into a living organism that works together to achieve the project's objectives while maintaining the individuality of its members. To achieve this synergy the PM must be the leader, mentor, problem solver, coach, and caregiver to the project team. This is no simple task and the number of publications available on leadership skills and techniques grows every day. The following represent the prominent techniques associated with building and maintaining an effective project team.

Lead by example: A PM can lead by example in the way he responds to challenges, conflicts, and success in front of the team. Discussions about a team member's performance should be held in private and with his or her supervisor engaged. A leader limits all negative and critical discussions to the parties directly involved in the situation. Team members will sense conflict and discord within the project and may look for alternative work before the project fails. A strong leader does not underestimate the value of recognition for a job well done. He gives credit to the performer and shares his satisfaction in public. Also, when the good leader makes a mistake in a public forum, the correction of his error is also made in a public forum. The PM applies these principles when addressing conflict between employees by avoiding the perception of taking sides and encouraging the parities to resolve the issue themselves.

> *Nearly all men can stand Adversity but if you want to test a man's Character, give him power.*
> —Abraham Lincoln

Communicate: Relationships are built and maintained through open lines of communication. This open line of communications is just as critical for team development as it was with resource acquisition. Open communication includes making the team aware of potential changes to the project or its team. Important information shared with the team as soon as is appropriate reduces rumors based on half-truths and exaggerations. Reducing rumors within the team goes a long way in developing cohesive teams.

Understand roles: This technique is equally important in developing teams. Respecting the roles and responsibilities between the PM, the assigned resource, and the functional manager will avoid strains on these relationships. An employee that feels unsupported or receives conflicting directions will look for other opportunities and frustrated managers will be reluctant to provide resources to future projects.

Be vested in the team: Establishing mutually beneficial relationships with members of the project team helps build commitment to the project. Each person on the team has unique career and personal aspirations. A PM can build lasting relationships by demonstrating interest in these aspirations. This knowledge will help the PM to identify ethical opportunities to aide in these pursuits within the bounds of the project constraints. The PM can help team members with work and home life balance by being aware of and respecting the personal commitments at home when scheduling meetings and travel.

Respect expertise: Every PM strives to have the best and the brightest people on their team. Desirable team members are passionate about their profession and strive to do valued work. A PM that seeks and uses a person's professional advice feeds the team members passion and drive.

> *It does not make sense to hire smart people and tell them what to do; we hire smart people so they can tell us what to do.*
> —Steve Jobs

Focus on requirements: People tend to excel when they have a clear understanding of the desired outcomes and they are given the latitude to apply their talents to achieve the objective. Valued performers gravitate to opportunities to contribute to the success using their expertise and quickly look for an exit when told how to do their job.

Grow talent from within: Every PM wants the best and brightest talent assigned to his projects. The reality is the best and brightest come in short supply. When trying to resolve resource shortfalls, an assumed solution is to hire additional people or to outsource the task. By going outside the team to solve the problem, the PM could be sending the signal that he does not see the potential within his team. By being vested in the career aspirations of each team member, the PM is aware of those willing to take on a new challenge that is mutually beneficial to the team and the employee.

Know the team: Many people have talents beyond their current position within the organization. In other cases, the talents required to do one type of work is like that of talents within another discipline. For example, test engineers often have experience with requirements analysis and user interface design, which are talents typically provided by system engineers and analysts. In other cases, a potential resource may aspire to learn a new talent. Valued team members appreciate the opportunity to apply all their talents to the project rather than being held to artificial boundaries.

Encourage teamwork: It is important to get highly qualified and talented people. But building and maintaining a cohesive team that collaborates and supports each other is equally important. A highly qualified technical expert that is incapable of working within a team may be less desirable than a highly ambitious and cooperative person who is less qualified.

Mind the gap: The term caregiver was used earlier to express the importance of ensuring the team is properly equipped to do their job. As the team comes together, people may not have the tools, computers, and facilities they require to perform. Employees do not care whose job it is to get them a computer, they only care that they do not have it. This last role of the PM is as important as all the others. Effective PMs are those that have the right people on their team and the team is equipped to execute.

The Tuckman team performance model: In 1966 Dr. Bruce Tuckman published a model that depicted the four stages in which a team goes through as it develops. The initial model included the four stages: forming, storming, norming, and performing. In 1975, he added a fifth stage for adjourning the team [1]. The following list of these stages is available from multiple sources including the *PMBOK Guide*.

- *Forming:* This stage is characterized by a high level of independence between the team members and varying degrees of understanding of the project scope and organization. The level of PM engagement in day-to-day activities is very high.

- *Storming:* At this stage, the level of understanding of project details begins to grow with members who are beginning to work independently on their assigned tasks. The need to foster teamwork and mutual respect is high at this stage as relationships are beginning to form. The need for the PM in day-to-day operational decisions reduces, freeing him to focus on planning and team development.

- *Norming*: Agreement and understanding on the scope and approach for the project becomes common. Partnerships within the team begin to form and trust is gained among the team. The PM now has a solidified team that is beginning to work in line with the project plan.

- *Performing:* The team is now performing at the desired level of interdependence, trust, and cooperation. The PM's attention is now free to address performance improvement, risk management, and team sustainment in a proactive manner.

- *Adjourning:* In this last stage the project has come to its conclusion and the resources must be released for assignment to other projects. The PM is focused on ensuring an orderly termination of work, capturing intellectual knowledge, and closing out charge accounts.

The key takeaway here is that teams go through a natural life cycle very much like that of a project. As with projects, each phase of the team's maturity does not come at the same speed and often must be repeated as change occurs in the project and the team. Successful project managers are aware of the current stage in which his team resides and monitors to ensure it progresses rather than declines in its maturity.

There is no single tip or technique that can help the new PM to form and develop successful project teams. However, there are a few common themes in these recommendations. Teams are successful when their members are challenged, their talents are valued, and they are treated with respect.

Manage the Project Team

Constant attention is required to ensure the team maintains the momentum achieved during its development. The goal of this process is to ensure resources are available to the project when needed and resource levels are maintained according to the work plan.

Monitor performance: As the project progresses through its work, performance indicators are generated by the schedule, cost, quality, risk, and configuration management activities. Each of these indicators is influenced by the resources assigned to the project. Cost and schedule variances can be an indication for needed change in the type or quantity of resources. An upswing in product rejects can be an indication of a need to review processes or training. A high number of change requests could indicate a need to review the performer's understanding of the work. By monitoring and understanding all performance indicators, the PM can be prepared to address potential resourcing issues.

Control responses: Performance indicators are only an indication of potential issues. The PM must proceed with caution when responding to challenges in front of the team. Discussions about performance indicators should be held in private with those directly involved in the situation. Root cause analysis is required to determine the source of any performance issue and the course of action required before one reacts to bad news.

Address issues: The tips and techniques provided thus far have leaned heavily toward the soft skills required by the PM. These soft skills are vitally important in acquiring, developing, and maintaining a team of resources. However, through all this the PM retains the right to expect quality work and on-time delivery. As the person accountable for project success, the PM has the right to address performance issues with the resource provider. When a subcontractor fails to meet agreed terms, the leadership of the contractor is to be contacted and advised of the unacceptable performance. The same expectation is in place between the PM and the manager of an internal resource. In both cases, the PM must communicate expectations, document failure to meet these expectations, and work with the resource provider to correct the issue.

Provide feedback: Particularly within matrix organizations, the PM often has more direct knowledge of a person's day-to-day job performance than their supervisor. Employees rely on positive feedback from PMs during performance reviews. A PM vested in the careers of his teammates ensures relevant information is shared with the appropriate supervisors at the right times during review periods.

Plan for attrition: Despite the PM's best efforts, people leave teams to pursue new challenges or to support higher priority projects within the organization. Another reality for the new PM is that when the organization faces a new challenge to its business or mission success, it too wants the best and brightest to solve the problem. There is a good chance that some of these highly desirable people are on the PM's team. The good news is that if the PM has been practicing the techniques previously recommended, he will understand the tasks on which the person is working and the skills required to replace this person. The PM will also know who within his team has these skills as well as those with a desire to build these skills.

A proactive PM also knows the single points of failure within the team. A single point of failure is when only one person understands the work and the plan to execute the work. These conditions represent risk to the project and an appropriate risk response is to cross-train others on the work and to perform periodic reviews of all documentation. In the event of the risk is realized, the PM must ensure that the departing person supports a full transition to the new resource.

Coordinate resource changes: Projects are time phased and resources move in and out of the project team as the schedule dictates. For example, a prototype shop cannot start working on the prototype until the initial requirements and concept have been defined. When finished with this task the shop may not be needed again until later in the project. Time will elapse between when initial agreements are established with resource providers and when the resources are scheduled to join the project. During this period resources may leave the provider's team, the functional manager may change, and the schedule itself is likely to change. Ensuring that resources show up to work when planned requires continual communications with the providers. Changes in the schedule may lengthen the time a resource is required for the project, which also must be coordinated with the provider.

Project teams are living organisms comprised of numerous individuals each with their own personal aspirations and lifechanging events. These resources are provided by organizations external to the PM and these organizations also experience change. Coupled with the potential for change within the project, the PM must focus attention on maintaining a healthy and productive team to minimize resource challenges.

Summary

Resource management is the set of activities performed by the PM to acquire, organize and enable the human and physical resources associated with the project. In order to successfully perform theses resource management activities, the PM requires a strong understanding of the soft skills used to recruit, motivate and retain the best available talent to the project.

Within the PM's organization there are roles and responsibilities associated with human resources, facilities and projects. How these responsibilities are allocated depends on the organizational construct of the PM's organization. Therefore, it is important for the PM to understand the type of organization in which his project resides and the parties that make decisions about how resources are assigned to projects.

The performance of project teams often follows a natural maturation cycle. The Tuckman model includes the stages of Forming, Storming, Norming, and Performing to describe the characteristics for the team's performance as it matures. The PM uses his soft skills to encourage partnerships and trusted relationships within the team. There is no single tip or technique that can help the new PM to form and develop successful project teams. However, teams are the most successful when their members are challenged, their talents are valued, and they are treated with respect.

As a living organism, the project team requires continual monitoring, nurturing and support to maintain momentum through to the project's completion. The PM uses performance indicators created in the schedule, configuration, vendor, quality, and risk management processes to identify potential issues. He continues to use his soft skills to determine the need to address a performance issue while treating the team with respect and understanding.

Throughout the project life cycle, people will join and leave the team due to competing demands, new opportunities and as the work plan dictates. Careful coordination with resource providers helps to ensure partnerships are maintained and people show up when needed. Loosing critical team members is a reality. The best and brightest are in high demand and they aspire to grown in their career. This reality necessitates the need to reduce single points of failure by cross training, knowledge sharing, and development of future star players.

References

[1] Project Management Institute, Inc., *A Guide to the Project Management Body of Knowledge (PMBOK Guide)*,Sixth Edition, Newtown Square, PA: Project Management Institute, Inc., 2017.

[2] Project Management Institute, Inc., *Managing Change in Organizations: A Practice Guide*, Project Management Institute, Inc., Newtown Square, PA: 2013.

CHAPTER 7

Vendor Management

A vendor is a third-party supplier that provides products or services. Vendor management includes the activities involved with the solicitation, acquisition, and oversight of vendors supporting a project. The goal of vendor management is to ensure that vendors performing contracted activities are integrated into the project team, that the vendors are positioned for success, and that all contracting policies and procedures are followed.

This chapter introduces the technical project manager to the processes, tools, and techniques associated with this process along with roles and responsibilities associated with acquiring and managing vendors.

Figure 7.1 depicts the relationship between vendor management and the other project management activities. The project manager is responsible for monitoring vendor performance, developing contract documents, ensuring multivendor integration, and maintaining vendor relationships. The remaining activities such as performing audits, risk management, documentation management, and project governance apply to all aspects of the project including vendor management.

While all these activities are important, ensuring the vendor's contract accurately reflects the scope, maintaining a strong relationship with the vendor, and monitoring contractor performance based on objective measurement criteria are critical to project success.

During vendor management the project manager works with the project team to identify elements of the project that may require outsourcing. The decision to outsource is often connected to the risk management process where an identified lack of internal capability, expertise, or capacity can be mitigated by using an external source. Early in the project life cycle, key stakeholders and subject-matter experts provide valuable insight into make-or-buy decisions regarding whether a product or service should be performed inside the organization or purchased from an external party [1]. Table 7.1 introduces the roles and responsibilities associated with the processes described within this chapter.

What gets measured, gets managed.
—Peter Drucker

The project manager relies on a procurement officer assigned to the project team to initiate and manage the contract. The project manager works with the project team to ensure the outsourced products and services are delivered in accordance with the contract specification. During the vendor selection phase, the

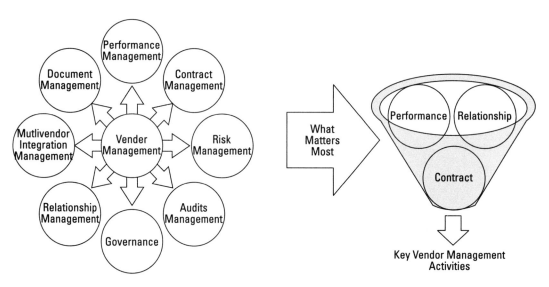

Figure 7.1 Vendor management.

Table 7.1 Roles within Vendor Management

Role	Responsibility
Customer lead	The member of customer's organization charged to manage the project from the customer's perspective. Has final authority over all additions, changes, or deletions to project scope, schedule, and performance.
Project manager	The person accountable for project success, the PM leads, monitors, and supports the vendor management processes and procedures.
Procurement officer	Member of the project manager's team that can authorize new agreements and changes to agreements between the project team and the customer or between the project team and its vendors.
Subject matter experts	Members of the PM's organization who are engineers, scientists, and analysts with expertise in one or more areas of the project.
Risk manager	A member of the project team with specialized training and skills in the identification and management of project risk.
Resource managers	Members of the PM's organization who develop and supply technical and project resources to a project team.
Work package lead	The person assigned responsibility and accountability to deliver a work package. May be a member of the PM's organization or the vendor organization if the work package is outsourced.
Quality control manager	A member of the project manager's team charged with the development and implementation of the quality control processes and data sets. Quality control is a process knowledge area that requires specialized training and experience. The PM should seek the support of a competent quality control manager.
Project scheduler	Member of the project manager's team that performs the consolidation, structuring, publication, and management of the project schedule. Role may be performed by the PM as a secondary function or by a dedicated resource.

project manager works with the project team to ensure the definition of clear requirements, terms, conditions, and key performance metrics for the outsourced effort. The PM also works with the project scheduler to ensure the project schedule is clearly communicated in the procurement documents. During negotiations with the selected vendor, it is imperative that these elements are agreed upon by both parties prior to issuance of a contract or purchase order.

Figure 7.2 demonstrates the typical activities associated with vendor management from the point of contract initiation through to payment for goods and services. While this diagram presents a logical sequence of events, it is important to recognize the recursive nature of these activities. Expect the need to revisit steps taken earlier in the process as new information is identified during the procurement life cycle.

Issuance of official changes to the contract after award are commonplace. However, greater attention paid to detail in the early steps will reduce the need for changes. A PM typically relies on a procurement officer to communicate with the vendor through the selection process. However, the PM must proactively monitor and support the process by ensuring the accuracy of all project-specific data and its inclusion in the contract documents.

The complexity associated with a procurement can range from purchasing COTS items to the engagement with a contractor to complete a turnkey design, build, and integration effort for the project. In the case of a COTS purchase, the effort may be as simple as issuing a request for quotation from one or more suppliers. At the other end of the spectrum is the competitive award of a turnkey design, development, integration, and final delivery of a complex system. Following the steps described in this chapter for all vendors regardless of cost and complexity will greatly improve the probability of success for any project. These steps will also help the project team, including vendors, to be aware of cost and schedule constraints while understanding their role in managing to these constraints.

During vendor selection, the PM works with the procurement officer to evaluate the offers received and to select the vendor best suited for the project. The key activities in the selection process are negotiation, past performance evaluation, and vendor selection. The goal of negotiation is not to achieve the lowest price but rather to achieve a win–win condition where both the buyer and the seller will benefit. When negotiating, the PM must ensure the supplier company can demonstrate

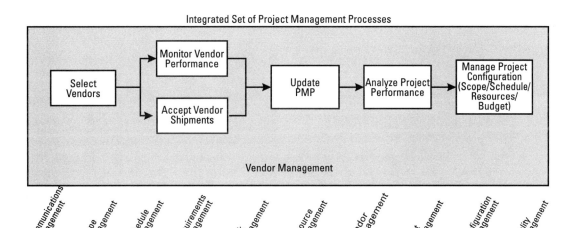

Figure 7.2 Vendor management activities.

the existence of internal standards and policies required to comply with the terms and conditions of the contract. Past performance data can be used to see if the company has a successful delivery record. Past performance databases such as a supplier corrective action request (SCAR) will indicate if the company has a history of poor performance reports and the ability to successfully recover when notified by the customer.

The vendor selection process results with the award of an official purchase order or contract, which are binding agreements between a supplier and a buyer involving the delivery of a product or service. These contractual agreements will include the contract type, scope of work, product requirements, payment method, quality management methods, schedule of delivery, and general terms and conditions.

The vendor performance process is used to identify potential issues in time to implement corrective actions. Supplier metrics should measure the contractor's performance in meeting cost, schedule, and performance criteria. The contractor's risk management reporting can also be an indicator of performance. The project manager should be involved in regular communications with the vendor and performance should be part of each discussion. Being actively involved in contractor meetings such as the project kickoff, periodic performance status reviews, and change control meetings are important for vendor relationship management.

Visit the vendor in person.
—Jerry Madden: NASA's Goddard Flight Center [2]

In preparation for vendor shipment it is important to ensure quality control processes are in place at the vendor site as well as at the project's receiving location. In many cases, particularly when the vendor is fabricating or integrating a product, quality inspections may be required prior to allowing shipment from the contractor facility. Similar processes should be used prior to acceptance of received products at the project facility. These processes should clearly define the methods and criteria used to inspect products, manage any noncompliance, and the terms for product acceptance. In cases where the product is to be integrated into a larger assembly or finished product, the acceptance process will continue until the end-to-end solution is accepted by the customer.

Vendor Selection

The objective in vendor selection is to identify the supplier that best demonstrates the capability and experience to deliver the required product or service within the cost, schedule, and performance constraints. The project manager works through the procurement officer to assess each offer, review each company's past performance record, and then negotiate final terms and conditions with the selected vendor. During the vendor selection process, it is important to concentrate on the value (cost, schedule, and performance) each vendor is bringing to the project. Concentrating solely on the lowest price will result in receiving the lowest quality of work.

There may be cases where the most qualified product manufacturer is not the most qualified product designer. For example; a manufacturer may find they can be more competitive and profitable if they focus on manufacturing and an engineering firm may find it is better suited to produce designs, leaving the manufacturing up to someone else. Therefore, one may find the product design is outsourced to a design agency with a different supplier providing the product according to someone else's design. This condition provides flexibility to both parties by allowing each seller and buyer to focus on core business. However, this hand-off between vendors can create the risk for gaps in the end-to-end scope. These cases increase the criticality of multivendor integration by the buyer's project team.

Figure 7.3 depicts the activity sequence for vendor selection. The process starts with the definition of the products and services required and the process ends with a contractual agreement between the buyer and the seller. In between these two steps, are activities associated with defining the strategy for procurement, selecting list of potential candidates, evaluation of their proposals, and final negotiations with the selected vendor.

Supplier requirements definition. The objective in supplier requirements definition is to clearly define *what* product or service is required, *when* the product or service must be delivered, and any specific conditions as to *how* the product or service is to be delivered. These details are included in a single set of procurement specification documents that include a statement of work, specifications for products supplied, and the terms and conditions associated with the contract. The project manager works with the project team, key stakeholders, and subject matter experts to define a clear set of requirements for the product or service.

The process and techniques for defining clear requirements are further described in Chapter 4.

The project scope of work document defines the scope, location of work, period of performance, deliverable(s), project schedule, applicable industry standards, acceptance criteria, and other special requirements. It is very important that nothing is open for interpretation. The procurement specification should be clear about what is required and expected from the supplier without telling the contractor how to do the work. Typical procurement specification data elements include

- Project scope and background;
- Detailed requirements (drawings, size, weight, performance parameters, safety requirements, sustainability requirements, etc.);
- Quality requirements and processes (acceptance criteria);
- Training requirements;
- Documentation required from supplier (test reports, factory acceptance test, performance reporting, first article reports, etc.);

Figure 7.3 Vendor selection steps.

- Transition requirements (e.g., does product needs to be transitioned in specific environment criteria);
- Service conditions and environmental factors;
- Delivery schedule;
- Other documents relevant to the requirement;
- Facility capacity (e.g., production within certain controlled environments);
- Quantities;
- Warranty and repairs (e.g., once product is accepted during system level test, warranty coverage).

Prescribing how the vendor is to complete their work shifts vendor risk to the buyer.

Develop procurement strategy. The objective of procurement strategy is to determine *how* the product and services will be acquired. The following information is used to develop a strategy:

- Type and duration of work to be performed;
- Level of uncertainty and risk associated with the work or product;
- Specialized skills, experience, or facilities required;
- Expertise required;
- Delivery schedule and any critical dependencies;
- Availability of funding;
- Any terms and conditions associated with the procurement.

The project team uses this information to determine the method to solicit offers for the procurement and the type of contract (cost plus, firmed fixed, or time and material) that fits these needs. Every organization has specific methods and policies for managing risk associated with contracting. Figure 7.4 depicts three of the most common contract types recognized in the project management community.

Fixed-price contracts are used when there is a well-defined statement of work, design, and specifications for the product or service. Fixed-price contracts are used when the work to be performed is a known element and significant industry experience exists. In a fixed-price contract, the seller agrees to perform the work at the negotiated contract value and the seller accepts much of the risk based on the details of the procurement specification. Fixed-price contracts place responsibility on the contractor to deliver the goods, services, or facility in accordance with the contract terms at a set price. The seller's price therefore includes the cost associated with all identified risks and the seller keeps these funds even if the risks are not realized.

In fixed-price contracting, the buyer shifts risk to the seller and pays the seller to accept the risk.

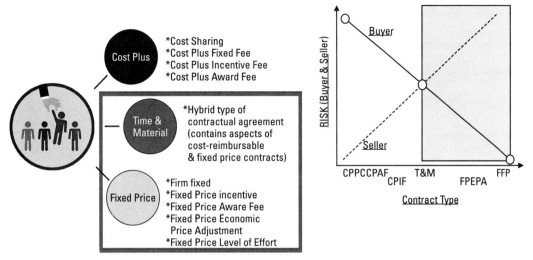

Figure 7.4 Contract types.

There will be occasions where information is not known at the time the procurement specification is created. When this condition occurs, the contractor may be due adjustments to the contract cost, schedule, or performance terms based on this new information. These modifications are often referred to as a change order. For this reason, the buyer is incentivized to ensure minimal opportunity for misinterpretation or assumptions by the seller. There are multiple types of fixed-price contracts:

- *Firm fixed price* (FFP). Most commonly used when the design and specification are very mature. Examples include build to design on full-scale development projects or full production projects. This is the simplest type, where all the terms are quite straightforward and easy to understand.
- *Fixed-price incentive* (FPI). Labor or material requirements are moderately uncertain, such as in the production of a major system based on a prototype.
- *Fixed-price award fee* (FPAF). An FFP with standards for evaluating performance and procedures for calculating a fee based on performance against the standards.
- *Fixed-price economic price adjustment* (FP-EPA). The stability of the market or labor conditions during an extended period of performance is uncertain and contingencies that would otherwise be included in the contract price can be identified and covered separately in the contract.
- *Fixed-price level of effort* (FP-LOE). The buyer agrees to pay a fixed dollar amount for a specified level of effort (hours) over a stated period of timed.

Buyer's failure to clearly define the Scope, requirements, acceptance criteria and contract terms creates opportunity for the seller to shift risk back to the buyer.

Time and materials contracts are useful in cases where there is urgency to receive services and it is prohibitive to wait until all aspects of the scope can be clearly defined. In these cases, the buyer can receive immediate services to respond to the urgency while allowing time to create a detailed scope. Under a time and material contracts, the buyer assumes all risk for cost, schedule, and performance. The contractor is not held to delivery or completion of a product and is required only minor justification for invoices. The use of time and materials (T&M) contracts is recommended only until the full scope can be defined and should be limited to a short period of performance.

Under a time and materials contract the buyer pays for supplies and services based on specified fixed hourly rates that include wages, overhead, general expenses, and profit. This type of contract allows for rapid delivery cycles. Once the buyer signs off on the vendor's estimate and statement of work (SOW), the work can start. A time and materials contract provides an agreed basis for payment for materials supplied and labor performed. Time and materials contracts should always have a cost ceiling to cap the allowable charges on the contract. Additionally, contractor invoicing and cost reporting must be timely and closely monitored by the PM to avoid runaway costs with little to no demonstrated results.

The buyer assumes all Risk on time and materials contracts and therefore should be used only when other methods will impact the customer's business or mission.

Cost-plus contracts are beneficial in cases where the specifications for the product or service are not fully defined and both the seller and the buyer recognize the need for flexibility in execution of the work. With cost-plus contracts, the buyer agrees to the seller's overhead and profit percentages but understands the quantity of the services and the cost of these services may require adjustment. Examples of this case include the acquisition of engineering and professional services or in the early stages of a project when concept development and technology evaluation are performed. The buyer agrees to pay the contractor for all expenses plus a fee representing seller overhead and profit. The contractor must fully document and report all costs to the buyer. The initial award value of cost-plus contracts is typically lower than fixed price contracts as the seller does not have to compensate for as much risk. With cost-plus contracts, the contractor is only required to put forth its best effort in the performance of the contract. Invoices are submitted to and paid by the buyer at an agreed frequency (monthly, quarterly) or at achievement of milestones. Depending on the size of the contract value, these contracts can be incrementally funded as project progresses.

- *Cost-plus-fixed-fee* (CPFF). Most commonly used for nonengineering projects such as exploratory projects and development projects. A fixed fee is defined at the beginning of the contract award; however, it may be adjusted due to changes in the work performed under the contract.
- *Cost-sharing.* Contractor is reimbursed for an agreed-upon portion of its allowable costs and agrees to absorb the rest of its cost. This method is useful in research and development projects where the seller retains the intellectual property due to their investment in the project.
- *Cost-plus-incentive-fee* (CPIF). Initial negotiated fee that be may be paid by the buyer if the contractor meets certain incentive criteria. Incentive criteria

can include realized cost savings or early achievement of milestones. The contractor is incentivized to control costs and meet performance objectives to realize a higher fee.

- *Cost-plus-award-fee* (CPAF). Provides for a fee that consists of a base fee amount (which may even be $0) that is negotiated prior to award, and an award fee amount, which is based on the buyer's subjective evaluation of contract performance. The award fee is designed to motivate excellence in the areas of cost, schedule, and technical performance. This contract requires the buyer to prepare an award fee plan that establishes the procedures for evaluating award-fee and an award-fee board for conducting the award-fee evaluation.

Vendor solicitation. With a clear definition of the required product and service, a recommended contract type, and an agreed upon procurement strategy the project team is now ready to engage potential vendors. The policies and procedures of the buying organization will define the process followed for solicitation. These policies will define methods used for requesting information, quotations, and proposals.

In some cases, the project team may already have a list of contractors identified that have the required capabilities and experience defined within the procurement specification. In this case, the procurement official may allow the team to go directly to request for quote (RFQ) or request for proposal (RFP).

In other cases, there may be a need to engage the vendor community to help identify those with the experience required for the project. The project team uses the request for information (RFI) process to establish the list of vendors to include in the RFP or RFQ.

An RFI is used by the project team to identify potential sources for the contract. An RFI requests input from a range of suppliers on their capability to meet the buyer's requirements and can help the organization understand if allowances should be made to create opportunities for small or minority-owned businesses. Sometimes an RFI can help the project team during the planning stages to determine if outsourcing is the best method to use for an aspect of the project. The RFI is used to obtain vendor feedback on the procurement specification prior to issuing a formal RFP or RFQ. Depending on local policies, the procurement official will use the RFI responses to narrow the list of vendors who will receive an official RFP or RFQ.

The RFQ is competitive bidding, usually used for COTS noncomplex products or service items where price is the most important selection criteria. RFQs are issued when the buyer has a firm understanding of the requirement and lower-level specifications. The RFQ process can also be utilized for production of known components, systems, or assemblies. The RFQ is used to determine who has the best price for a well-defined procurement.

The RFP is used when the buyer requires a balance of capability, experience, and cost from the sellers. RFPs are used when there is a lack of a predefined or assumed solution to the project requirements. Venders are asked to propose a solution based on their understanding of the project requirements. Unlike the RFQ, where an equipment list is provided, the vendor is expected use their expertise to identify the labor, equipment, and materials required to deliver their proposed solution. The vendor will respond by using their expertise to create a proposal with their best ideas, solutions, and price.

The level of formality in an RFP is significantly greater than the RFI and RFQ because the buyer must communicate a clear set of expectations for the project and the vendor must clearly demonstrate their understanding of the buyer's needs. The vendor's proposal communicates how their proposed solution best fits that need.

Items not specific and clear in the RFP leave the vendor open to change orders after the contract award.

Vendor solicitation is started early in a product development phase to support project timelines. The time frame required to create a procurement specification, soliciting proposals, evaluating proposals, selecting the vendor, and awarding the contract can range from several months to several years. The RFP outlines technical, administrative, and financial information for vendors to submit proposals.

For any RFI, RFP, or RFQ, the following criteria should be requested from the vendors to aid in evaluation of their proposals:

- Introduction and executive summary;
- Vendor business overview and background;
- Recognition of project objectives, page count, contract type, statement of work and schedule information, system overview and requirements, assumptions and constrains, terms and conditions, budget, references, and selection criteria;
- Vendor's years of experience;
- Number of similar projects performed;
- Qualified personnel assigned to the project.

Evaluate vendor responses. The objective when evaluating vendor proposals is to identify the offeror that is most likely to meet the project scope and performance requirements within the project's budget according the project's schedule. During contract requirements definition, assessment criteria were defined for the evaluation of each vendor proposal. These criteria allow each proposal to be assessed based on qualifications, ability to satisfy the requirements, and the timeline to deliver the scope before evaluating the proposed cost. This approach reduces the temptation to focus on the lowest cost proposal that may not meet the scope and expectations of the project. Figure 7.5 provides example criteria for objective analysis of each offeror's ability to meet to the project's cost, schedule, and performance constraints.

Draw a clear distinction between what is needed versus what is desired. The criteria should differentiate one vendor as the one most likely to deliver the product or service successfully. Use criteria that directly trace to the project requirements. Review and rate each vendor proposal based on their technical, management, past performance, and price.

Criteria for evaluation of proposed solutions (performance) is based solely on requirements.

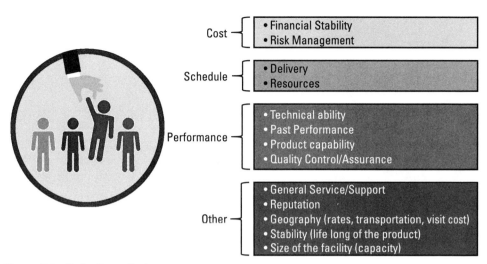

Figure 7.5 Evaluation criteria.

Each customer places a unique level of importance on the project's cost, schedule, and performance. In one case the customer may be willing to pay a higher price to get the project completed sooner and in another case the customer may be willing to accept a less-than-perfect solution if it saves time and money. Under requirements management, the project team helped the customer to assign priorities to each requirement in the event every requirement could not be met or afforded. Use these customer expectations to assign weights to the proposal evaluation criteria. These weighted criteria help the project manager to balance trade-offs among the cost, schedule, and performance parameters.

In the first example where the customer was willing to pay for faster delivery, cost may be assigned a lower weight as compared to performance and schedule. In the second example, certain features of the solution may receive lower weights and cost may have a higher weight. The idea with weighting is to create objective criteria based on customer priorities. As depicted in Figure 7.6, weighting helps the project team and the customer to make informed decisions based on the customer's mission or business priorities.

The expected outcome of this step is a list of offerors narrowed down to the two or three candidates worthy of further consideration. However, this is only the first half of the vendor selection process. Just as one would not buy a new car without first driving it, one should not select a vendor based on a sterile written proposal.

The next step is to verify that the vendor can deliver as promised. Techniques such as system demonstrations, side-by-side comparisons, interviews, prior projects reviews, and negotiations can help to validate the vendor's proposal.

Vendor negotiations. The objective during vendor negotiations is to minimize any remaining uncertainty in the offers and identify the best candidate for award of the contract. This point will likely be the first time the project management team can directly communicate with the vendors about their specific proposals. This process can provide each of the remaining vendors the opportunity to clarify or amplify their offers. Extreme care should be taken to ensure all remaining vendors are given the same opportunity. Establish good two-way communication with the

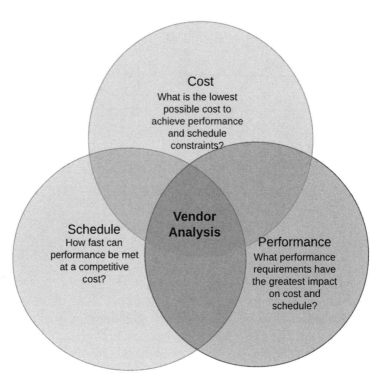

Figure 7.6 Balance of cost, schedule, and performance.

selected two or three vendors before starting the negotiation process. The objective is to review the vendor proposals and conduct detailed negotiations on product cost, schedule, and performance constraints. During negotiations be honest, ethical, and fair. avoiding human emotion and political positioning. Maintain a win-win strategy for both parties throughout negotiation. Remember, do not drain every cent out of the vendor for the lowest price during negotiation; this approach establishes a poor foundation for executing the project. The negotiation is the start of a partnership in which the buyer and the seller work together to deliver a solution that meets the cost, schedule, and performance constraints of the customer.

Use negotiations to validate detailed aspects of the solution while maintaining sensitivity to high-risk areas identified during risk management. Remember, the goal is to minimize uncertainty while establishing a strong relationship. Be careful not to allow the project scope and cost to grow during negotiations. It is an easy temptation to accept seller upgrades that do not relate directly to the customer's requirements or directly reduce overall project risk. Add any changes, clarifications, assumptions, and agreements made during negotiations into the official contract documentation. Additionally, the PM should be sure to understand and communicate her contractual authority (or lack of) to all vendors involved in the negotiation process.

Award contract or purchase order. The process, policies, and authorities for procurement vary between organizations. One may be surprised to learn that the PM has only an advisory role in his agency's procurement process. In some cases, the customer may also have a role in the approval of procurements. It is imperative that every project manager understands theses organizational processes and ensures that an authorized procurement official is assigned to the project as the start. It is never too early to engage with the procurement official. Be sure to hold early and frequent discussions about planned procurements with the assigned representative. Regardless of the process, a vendor is selected and awarded a contract or purchase order. At this time, ensure all contractual documents are clear and in place and that all documentation has been updated to reflect the results of the proposal evaluation and negotiations. Verify that the following key information is located and accurate:

- Supplier statement of work and/or procurement specification, including associated technical documents;
- Contract type (fixed price, cost plus, and/or time and material);
- Contract terms and conditions;
- Supplier on company tracked approved vendors list;
- Contract value;
- Payment method;
- Delivery schedule;
- Assumptions and constraints;
- Nondisclosure agreement;
- Property data rights.

When placing a vendor purchase order, the project manager reviews the contract document with the finance lead, quality lead, and technical lead before the procurement official signs the contract. Review all documents, including

- Contract statement of work;
- Procurement specification (associated detailed documents if required);
- Contract type;
- Payment processes, systems, and methods;
- Assumptions and constraints;
- Contract data requirements (monthly status, technical document, management plans, etc.);
- Delivery schedule;
- Property data rights;
- Signed nondisclosure agreement;
- Signed proprietary information agreement (if required).

For the purchase of COTS items or a blanket purchase order, it is important to work with the procurement official to identify basic parts information required for the purchase order, including

- Quantity order;
- Part number;
- Description;

- Unit price;
- Delivery schedule;
- Quality requirements (certificate of conformance, test report, etc.)

Vendor Product/Service Acceptance

Vendor product/service acceptance involves the receipt of vendor deliverables, acceptance of the deliverables, and authorizing payments to the vendor. The objective of vendor product acceptance is to ensure that vendor deliverables are consistent with the specifications defined in the contractual agreement and prompt payment is made to the vendor. In addition to physical goods, vendor deliverables includes services, documents, designs, software, or other tangible items received through the contract action. Therefore, the process for acceptance will vary depending on the deliverable and the contract requirements. The generic process for vendor product/service acceptance is depicted in Figure 7.7. In this process, the core steps are shown in bold along with the potential variations described herein.

During the core process, the vendor delivers a product or services authorized by the contract agreement. The product or service is accepted by the buyer who uses some sort of quality control methods to confirm the deliverable meets expectations and then the buyer pays the seller for the product or service. The exact quality processes are defined within the contract agreement, which includes the process for deliverable submission, the methods and metrics for approving or rejecting deliverables, and how issues with deliverables are adjudicated.

Product/service delivery. Throughout the project, the vendor will complete tasks that produce a deliverable specified by the contract. If the contract requires a predelivery approval, the buyer and seller follow the quality assurance methods defined within the contract. Examples of quality control measures may include

- Review and feedback on design artifacts and written reports;
- Completeness of equipment or product order;
- Inspection of products prior to shipment;

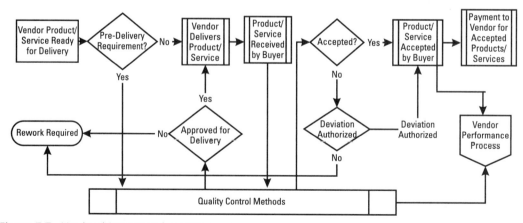

Figure 7.7 Vendor shipment and acceptance process.

- Testing of products prior to shipment;
- Review of vendor quality assurance audits.

If the product fails to meet the quality control metrics, the vendor's request to delivery will most likely be rejected until the shortfalls are addressed. The results of each quality control process are documented in the vendor performance database maintained by the buyer's quality manager.

Upon approval to deliver the product is shipped or transmitted to the buyer's receiving location. As the material makeup of deliverables can vary, the location can be a warehouse, project site, electronic repository, or a specific person.

In cases where the deliverable is a service it is equally important to have predelivery conditions in place. For example, a painting contractor should not show up at the jobsite until the services are required, or a network administrator should not show up and start billing until he has access to the network.

Product/service received. Each time a delivery is completed, the buyer performs steps to verify the delivery meets contractual requirements. The contract's quality control measures are used to document deliverables received, to validate compliance with the contract, and to approve or re-reject the deliverable. Examples of quality control measures may include

- Final review of design artifacts and written reports;
- Accuracy of equipment or product shipments;
- Inspection or testing of products received;
- Verification of services rendered;
- Inspection for loss or damage during shipment;
- Testing of products received;
- Review of vendor quality assurance audits.

If the deliverable fails to meet the quality control metrics, the vendor's delivery will most likely be rejected until the shortfalls are addressed. Rejection does not always mean returning the product to the vendor. Rejection can involve vendor corrective measures that include onsite correction, shipping of missing products, or vendor retrieving the rejected deliverable. There are also cases where the buyer elects to accept a deviation in the deliverable until corrections can be remedied. The results of each quality control process are documented in the vendor performance database maintained by the buyer.

In cases where the deliverable is a service, it is equally important to have quality control conditions in place to ensure the vendor provided the service and that the service meets the performance requirements of the contract.

Product/service acceptance. Upon acceptance of the deliverable, the product is logged within the buyer's inventory control system and utilized by the project. Notice of product acceptance is transmitted to the vendor and the appropriate elements of the project manager's organization. Acceptance of a deliverable means achievement of a project milestone that requires input to the project schedule, status reports, and progress metrics. Additionally, achievement of this milestone may trigger an invoice to the project manager's customer.

Quality control methods play an important role in product/service acceptance as they validate that the vendor has met the contractual requirements for each product or service. Quality control is a process knowledge area that requires specialized training and experience and the PM should seek the support of a competent quality control manager. Quality control is accomplished through product inspection, noncompliance tracking, and the management of noncompliance deviation requests. A noncompliance deviation request is the mechanism by which a vendor can ship a product with a known fault based on a plan to correct the shortfall later. The vendor submits a noncompliance deviation request, which must be approved by the buyer prior to shipping the product. With deviation requests, the quality control manager requires the supplier to provide root cause corrective action at the time of request. All noncompliance issues are tracked using SCARs, all of which are tracked in the project's quality management data set.

There may be an occasion where corrective action is not achievable. This condition is most likely to occur during development of new technologies where a performance parameter may not be achievable within cost and schedule constraints. In these cases, the project manager and the quality control manager may work with the customer and procurement officials to allow the shipment if there is no impact to safety. In the gas station use case the ability to accept a different grade of gasoline may be acceptable if it meets the minimum requirements for the vehicle. However, if the vehicle requires a high-octane fuel to operate properly, the PM and quality manager (QM) will not accept a low grade of fuel. In all deviation cases the contract documentation must be updated to reflect the change in acceptance criteria (the requirements). Detailed records management is imperative during quality control and product acceptance. The following information should be recorded for all shipments:

- Vendor shipment;
- Receipt of shipments (first article inspection, factory acceptance test, and/or common incoming quality inspection);
- Product acceptance;
- Transfer product to actual location (specific building or lab to integrate product into higher-level system).

Many manufacturing companies have quality control systems in place that are compliant with specific industry standards. The buyer may elect to accept the vendor's quality control system as the preshipment inspection process. In these cases, preshipment quality inspections are by the seller. Once a product is received at the buyer facility, the buyer inspects for basic quality prior to accepting the product.

The International Organization for Standards (ISO) is an independent, nongovernment body that publishes 22,000 standards used by almost every industry. The ISO 9000 family of quality management systems provide organizations with a known set of tools and processes for delivering products and services to customers. ISO 9000 is an example of a vendor provided quality control method that can meet the buyer's quality needs [2].

Basic quality inspections often apply to standard parts, electronic components, and COTS items. Accepting the vendor's quality management system (ISO 9000) does not eliminate the need to inspect supplier shipments. Therefore, the PM and QM institute safeguards to ensure that shipments adhere to quality specifications. Safeguards may include testing of a sample population for each shipment, detailed failure logging, and/or postacceptance return reporting. When a product has a high rate of failures during internal basic quality inspections or the postacceptance records indicate a potential issue, the project manager works with the quality lead and vendor to address the issue. Resolution of quality control gaps may require the addition of new quality inspection before vendor shipment. The complexity of a product or service drives the level of complexity for the quality control measures. For example, complex product or turnkey system integration may require a factory acceptance test (FAT) to verify that the integrated products work as specified before the vendor ships the product. FATs can be as detailed as the final acceptance testing but are held within a controlled environment and may only include a sampling of the components or subsystems. Therefore, FATs are not a replacement for full system testing within the operational environment as performed during final acceptance testing.

> *Factory acceptance testing is a hybrid form of developmental and acceptance testing in which the complete solution is tested to demonstrate the functionality and performance of the integration prior to deployment.*

The review and acceptance of products and services provided by consultants are usually performed by subject-matter experts within the buyer's project team. Designs and specifications often require multiple comment adjudication steps before official acceptance. The approval criteria for engineering and planning documents can sometimes be subjective due to the specialized nature of the work. Insisting on detailed documentation of all comments and corrections provided to the consultant is critical to evaluating performance. These logs not only provide invaluable data for dispute resolution, they also inject formality and clarity into the vendor communications.

Quality control is not limited to the shipping process. Development of a new product often includes the development of new manufacturing processes and mechanisms. One does not want to wait until 10,000 units have been produced to determine if the process delivers an acceptable product. New production processes go through initial quality control steps prior to shifting to full-scale production. Methods such as low-rate initial production (LRiP) and first article inspection (FAI) provide safeguards to test the process, mechanisms, and the resulting product prior to full-scale production.

Avoid the temptation to allow the schedule and the desire for demonstrated results to take priority over quality. Allowing a product to ship that is later rejected by the customer causes rework and creates far more impact on the project schedule than any delay to ensure problems are resolved early in the process. First article inspection data from the supplier includes

- Certificate of conformance;
- Test reports of subassemblies and fabricated parts;

- Noncompliant deviation approval record;
- Thermal and structural analysis reports, if required;
- System, subsystems, and competent-level test reports, if required;
- Requirements verification traceability metrics to show validation of all requirements;
- Special tooling and equipment process record.

Project quality management is further explored in Chapter 10.

Payment to vendor. An efficient vendor typically invoices for products as soon as they are transmitted from their facility. Payments to invoices typically require a confirmation from the project management office, which is supported through the product/service acceptance process. However, with the advent of electronic transactions, invoices are often paid on receipt unless otherwise directed by the PM. For example, the U.S. Department of Defense automatically pays all invoices within a period of days after receipt, leaving it up to the PM to file a rejection to recover paid funds. The PM must be aware of the invoicing and payment procedures and information systems used by her organization to avoid payment for unacceptable deliverables. The vendor's motivation to solve issues may quickly decline after receipt of payment.

Monitor Vendor Performance

The objective of monitoring vendor performance is to establish and maintain a collaborative relationship with the vendor in which risk and issues are communicated before significantly impacting the project's cost, schedule, or performance. This relationship should be based on open communications and objective performance criteria. Key activities associated with vendor performance are to manage the relationship, monitor performance, understand vendor risks, and manage changes to the contract.

Vendor relationship management is essential to project success. Within one month of contract award, schedule a project kickoff with the vendor at their facility. Include as many members of the team as possible in this kickoff meeting. During this meeting review project scope, requirements, schedule, risks and delivery expectations. Set expectations for periodic (weekly or biweekly) verbal communications to review progress, risks, and issues.

Being available for both formal and informal direct communications with the supplier will help to build and maintain a strong relationship in which the supplier is comfortable sharing challenges and concerns. An example of a healthy relationship is one where the supplier contacts the project manager about an issue and together, they discuss and agree on the recovery plan.

No project plan is perfect, and all projects face some sort of challenge. By maintaining a trusting, open, and collaborative relationship with each vendor the impact of these eventualities is greatly reduced. The need for requirements or scope change is also inevitable. The key to successful project management is to minimize avoidable change while controlling the impact of unavoidable change on the project. Avoidable changes are those items that could have been identified and addressed prior to contract award. Change impact can be minimized by timely identification and management of the need for change. A collaborative relationship with the vendor will help facilitate the change process and reduced the impact on overall project cost, schedule, and performance.

Vendor performance monitoring is the formal communication mechanism in which the vendor reports against a defined set of criteria. A clear set of expectations for tracking vendor performance was developed during the procurement planning phase. These reporting requirements were included in the procurement specification as part of the documentation required section. Examples of performance reporting include

- Schedule progress (on time/variances);
- Actual costs incurred (as-planned/variances);
- Risk register updates;
- Major accomplishments (recent);
- Planned work (look ahead);
- Quality of work delivered (accepted, rejected, corrected);
- Customer relationship (customer service, timeliness, responsiveness).

Avoid the temptation to ask for reporting that does not provide value to the project. The vendor passes the cost of reporting to the buyer. Reporting purchased and not used is waste and can have a negative impact on the vendor relationship.

An unread report is the equivalent of an unreturned voice mail or email.

Reporting is not a replacement for verbal communications and collaboration. The reporting data guides the focus and agenda for the periodic meetings. During these meetings review the reported data (quality, delivery, cost, customer service) to identify potential issues and develop a corrective action plan. Ensure that each action plan has an owner (supplier and buyer), a due date, and a process to clearly communicate progress between both parties. Items identified during these meetings become agenda items for following meetings.

Scorecards and dashboards provide the ability to report performance in a format that draws attention to the data points that directly indicate vendor performance. This method helps to reduce the level of detail reported while increasing buyer comprehension. Figure 7.8 provides an example scorecard for reporting cost, schedule, quality, and customer relations with shading used to indicate good and poor performance.

In this example, the scorecard is a roll-up of more detailed information that can be referenced during discussions with the vendor. Pay close attention to the scorecard and engage the vendor to understand performance impacts.

In the event that the vendor faces a performance challenge, engage to understand the root cause of the challenge and establish a get-well plan. Work closely with engineering, subject-matter experts, and the quality team to help the vendor to develop recovery strategies and to prioritize actions. Maintaining professionalism with the project team and the vendor when addressing critical issues is an

Categories	Description	Score
Quality	· # of SCARs · SCAR response time · SCAR resolution time · Past due SCARs · Weighted Average defect density	Poor
Delivery	· On-time delivery · Schedule variance	Good
Cost	· On-cost delivery · Cost Variance	Poor
Customer Service	· Get to green plans · Responsiveness · Overall communications	Marginal

Figure 7.8   Sample scorecard.

imperative. The project's success relies on the entire team and if one part of the team fails, the project fails. Be proactive and persistent in responding to vendor issues. If necessary, increase the frequency of meetings and status reports to keep the issue in the forefront. In cases of urgency, the PM and the procurement team may help the vendor to identify a subsupplier that can expedite elements of the vendor's scope to recover from late deliveries.

Do not assume the vendor is solely responsible for the shortfall. Root cause analysis may indicate a lack of clarity in the project scope and requirements or the issue may be the result of the buyer failing to deliver its contractual obligations to the seller. At the end of the day, every company is in business to be profitable and most companies use growth to maintain profitability.

Project failure is the last thing a company desires.

Vendors on occasion fail despite best efforts and collaboration. In this worst-case situation, be prepared to terminate the vendor and line up a new vendor. Termination of any contract leaves a bad reputation and huge impact to cost and schedule. Ensure every option has been exhausted before deciding to cancel the contract due to failure to meet contractual obligations. Avoid selecting a vendor that is single-point failure; keep all options open in the case a new supplier is needed to complete the work.

Contracts management involves the actions required to maintain the accuracy and structure of the procurement specification, contract documents, and purchase orders. As discussed earlier, change is inevitable on every project. The objective of contracts management is to ensure only necessary changes are authorized, all changes are documented, and only duly appointed officials authorize such changes. Project managers are frequently challenged with the unauthorized direction to the vendor for changes in scope, quality, or schedule. For example, a member of the engineering team suggests to the vendor that an increase in a performance requirement is required. This suggestion is interpreted by the vendor as an authorized change. The change is implemented and an invoice for the cost is generated. The engineer and the vendor did not engage the proper officials to approve and document the change and the buyer is now obligated to pay for the unauthorized change.

Summary

Vendor management is an ongoing process led by the project manager from the beginning of the project until all products have been delivered, accepted, and invoices have been paid. The project manager's key responsibility is to be transparent, to ensure requirements flow from customer to the supplier, and to maintain up-to-date, accurate contractual documentation throughout project life cycle. The PM relies on the expertise and authority of his procurement official throughout the project.

The project manager ensures that all changes are clearly communicated with key stakeholders, the project team, customer, supplier, and anyone that could be impacted by the change. Key activities include, outsourcing decisions, development of procurement requirements and specifications, determination of contract type, vendor selection, contractual agreement, tracking the key performance metrics, evaluating results, product or service acceptance, and payment.

The project manager ensures that all decisions about outsourcing, vendor selection, and vendor performance are communicated to the project team, stakeholders, and the customer. Understanding roles, responsibilities, and authorities associated with vendor management is imperative for the entire project team and will serve to avoid strains on the vendor relationship.

Very few projects are identical, and each requires a unique procurement strategy based on the type of work, level of risk, degree of uncertainty, and the overall cost, schedule, and performance constraints. The procurements strategy identifies the best mix of in-house and outsourced talent, the contract type for each vendor solicitation, and the type of contract best suited for each vendor. Engagement with an authorized procurement official early in the process will reduce errors in judgment and delays in the schedule.

Vendor solicitation and selection requires formality and clear documentation. Solicitation documentation minimizes the opportunity for interpretations and omissions during the proposal process. Scope statements, requirements, terms, and conditions require input from a broad range of expertise. Rely on the project team, stakeholders, subject-matter experts, and even the vendor community when developing a procurement specification.

Vendor selection is based on objective evaluation of all proposals, maximizing fairness and competition. The vendor selected is the one that best demonstrates the capability and experience to deliver the required product or service within cost and schedule constraints.

After award of the contract, the project manager maintains a mutually beneficial, trusting, and collaborative relationship between the project team and the vendor. This relationship requires open and responsive communications between the PM and the vendor. Reporting and submissions from the vendor must be used by the project team and all concerns should be discussed directly with the vendor.

Quality control is a critical tool for managing vendor performance. The project's quality lead ensures inspections are performed before and after the shipment of products. Each product or service requires a different quality control processes and criteria. The results of the quality control process for all products and services must be thoroughly documented. Quality control continues for the duration of the project and requires a strong relationship with the vendor.

Project success depends on the success of the entire project team. The team is composed of internal resources, and contracted vendors, and suppliers. The success of vendors and suppliers is equally important to project success and is the primary objective of vendor management.

References

[1]   *A Guide to the Project Management Body of Knowledge: PMBOK Guide*, Sixth Edition, Newtown Square, PA: Project Management Institute, Inc., 2017.

[2]   Maden, J.,and R. Stewart, "One Hundred Rules for NASA Project Managers," July 9, 1996 [online], https://www.projectsmart.co.uk/white-papers/100-rules-for-nasa-project-managers.pdf.

[3]   International Organization for Standardization (ISO), *ISO 9001:2015 Quality Management Systems*, International Organization for Standardization, Geneva, 2015.

CHAPTER 8

Cost Management

As the project schedule represents the time phased plan to deliver the products and services associated with a chartered work effort, the project budget is the time phased allocation of resources to the activities within the schedule. The project cost management process involves the activities to establish, monitor, and control the execution of the project budget.

In schedule management, progress is monitored by establishing a baseline plan then tracking progress against that plan. Schedule variances provide snapshot indications to the project's schedule performance. In cost management, the PM uses the variances between the planned cost and actual cost to obtain snapshot indicators of the execution of the budget. However, in both cases the PM's awareness is based on separate indicators for cost and schedule created with historic data. This method of project monitoring and control leaves many open questions:

- Is the overall project on, under, or over budget?
- For each work package, is effort on, under, or over budget?
- Does the rate of work completion support the rate of expenditures?
- What impact does the schedule have on the budget?
- How much will it cost to complete the project based on current progress?
- How long will it take to finish the project based on current progress?
- Are the performance results for this month part of a trend?
- What is the projected trend for the budget and schedule if things do not change?
- If the project is over budget or behind schedule, can it recover?
- How well must the project perform to stay on schedule and budget based on the current progress?

Project risk is any uncertain event or condition that could impact the project's ability to meet its objectives within cost, schedule, and performance constraints. The above questions are all example information points used to monitor for and detect risk events within a project.

An earned value management system (EVMS) is a project monitoring and control method that allows the PM to answer these questions thus establishing a higher level of insight to the project's ability to meet its cost, schedule, and performance objectives (see Figure 8.1). Earned value management integrates cost and schedule performance and provides the ability to forecast future performance based on historic trends (see Figure 8.2). Unfortunately, these benefits are often not enough to motivate project teams to adopt an EVMS. EVMS is often considered too complicated and costly to implement.

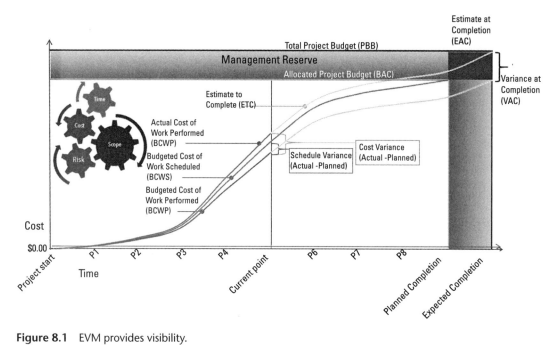

Figure 8.1   EVM provides visibility.

Acronym	Description	Formula	What does it mean
BAC	Budget at completion	Project Budget	Total budget of the project
AC or ACWP	Actual cost or actual cost of work performed	Actual Cost spent	What you already spent on the project
EV or BCWP	Earned value or budgeted cost of work performed	BAC * Actual % Complete	What you accomplished
PV or BCWS	Performed value or budgeted cost of work scheduled	BAC * Planned % Complete	What you planned to spend on the project
Key Performance Variance			
CV	Cost variance	EV – AC	Am I over or under budget
CPI	Cost performance index	EV / AC	>= 1 is good. 1 is on target (normalized)
SV	Schedule variance	EV – PV	Am I behind or ahead of schedule
SPI	Schedule performance index	EV / PV	>=1 is good. 1 is on target (normalized)
Performance Variance			
EAC	Estimate at completion	EAC + AC	what will be spend on whole project?
ETC	Estimate to completion	BAC – EV	what will be spent on remaining project?
TCPI	To complete performance index	(BAC – EV)/ (BAC – AC)	How well our project must perform to stay on budget
VAC	Variance at completion	BAC – EAC	Variance of total project cost from budget

Figure 8.2   Principle EVMS indicators.

This chapter will introduce the technical project manager to the fundamentals of EVMS while dispelling many of the misconceptions often associated with EVMS. The following resources are available to the PM to use in leading the implementation of an EVMS for her project.

ANSI.EIA Standard 748:c Earned Value Management Systems. American National Standards Institute (ANSI) and the Electronics Industry Association (EIA). Developed for the U.S. Department of Defense, this guide is widely accepted by U.S. Government and Commercial organizations. The standard is comprised of 32 individual guidelines for a successful earned value management system [1].

At first glance one may think all the elements defined in the EIA 2015 are to be completed in addition to the project management activities. However, many of the 32 elements shown in Figure 8.3 are included within this document as common-sense project management practices. Lastly, the 32 elements are the recommended practices by which an organization can validate the maturity of their EVMS. The use of all 32 is only required when the project's scope of work specifies the use of a compliant EVMS. In all other cases, it is up to the project manager and her lead-

Organization	Planning, Scheduling and Budgeting	Accounting Considerations	Analysis and Management Reports	Revisions and Data Maintenance
1. Define Work Scope (WBS)	6. Schedule with Network Logic	16. Record Direct Costs	22. Calculate Schedule Variance and Cost Variance	28. Incorporate Changes in a Timely Manner
2. Define Project Organization	7. Set Measurement Indicators	17. Summarize Direct Costs by WBS Elements	23. Identify Significant Variances for Analysis	29 Reconcile Current to Prior Budgets
3. Integrate Processes	8. Establish Budgets for Authorized Work	18. Summarize Direct Costs by OBS Elements	24. Analyze Indirect Cost Variances	30. Control Retroactive Changes
4. Identify Overhead Management	9. Budget by Cost Elements	19. Record/Allocate Indirect Costs	25. Summarize Information for Management	31. Prevent Unauthorized Revisions
5. Integrate WBS/OBS to Create Control Accounts	10. Create Work Packages/ Planning Packages	20. Identify Unit and Lot Costs	26. Implement Corrective Actions	32. Document PMB Changes
	11. Sum Detail Budgets to Control Account	21. Track and Report Material Costs and Quantities	27. Revise Estimate at Completion (EAC)	
	12. LOE Planning and Control			
	13. Establish Overhead Budgets			
	14. Identify Management Reserve and Undistributed Budget			
	15. Reconcile to Target Cost Goal			

Figure 8.3 ANSI/EIA-748-C 32 guidelines [1].

ership to determine the tailored approach for using earned value management to manage the individual project.

It should also be noted that ANSI/EIA 748 contains a set of practices for an earned value management system and does not directly represent a process. However, the PMI *Practice Standard for Earned Value Management* [2] does present a process model consistent with the practices described in the ANSI/EIA 748. Figure 8.4 represents the sequence in which activities associated with establishing and using an EVMS for cost management are presented in this chapter. These simplified recommendations are consistent with those described both within ANSI/EIA 741 and the PMI *Practice Standard* for EVM.

Figure 8.4 also conveys the message that cost management with or without EVMS is one of a set of integrated project management processes widely adopted by project organizations and described within this publication. Each of these processes and their outcomes are utilized and dependent on the other processes. The adoption of an effective EVMS is enabled by the project management processes at the bottom of Figure 8.4. Within each of the following sections, the activities covered elsewhere within this book are identified as such with new content focusing on the PM practices not specifically described elsewhere within.

The History of EVMS
The U.S. DoD has used an integration of cost and schedule data to manage its major acquisition programs since the 1960s. The DoD developed Cost/Schedule Control Systems Criteria (C/SCSC), which contained 35 elements and was used through the 1970s and the early 1990s as an element of the DoD Guide for Acquisition Management (DoD 5000i). In 1996, the DoD looked to industry to develop and manage its cost management standard. The resulting ANSI/EIA-748 Standard for Earned Value Management Systems is now in its third revision containing 32 guidelines. In 2005, U.S. Office of Management and Budget (OMB) implemented EVMS on all federal agency capital asset acquisitions and interagency projects that met a certain budget level. Today, EVMS is utilized by a wide range of commercial and industrial organizations to improve the rate of project success.

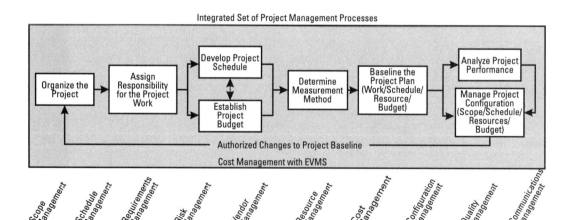

Figure 8.4   Cost management with EVMS process.

Historically, budget expenditures were monitored separately from project schedules. In this model, the PM was aware of how much she was spending, but could not correlate the rate of expenditures to the progression of work. The key difference between tracking progress utilizing traditional cost management methods versus an EVMS is the ability to link expenditures to work progress. EVMS assigns a budget line to each unit of work based on the planned cost and duration of the unit of work. The budget value for each element of work is earned as it is completed. The cost of the work performed reflects the total resources used to complete the task. In traditional cost management, the costs were often tracked by labor category or by invoices from the vendor with little linkage to what was accomplished with the resources. With EVMS, the PM knows what work was completed, the budgeted value of the work, and how much was spent to complete the work. The PM can use this information to predict the future performance of the project.

Table 8.1 provides the roles commonly associated with cost management and EVMS.

Organize the Project

Define work scope (WBS). This process aligns to the steps described in Chapters 2 and 3. The objective of this process is to decompose work down to discrete elements of work that can be assigned to a single resource group. In EVMS the WBS is the common schema to which all task, duration, schedule, and budget data are aligned. The WBS demonstrated in schedule management is provided in Figure 8.5.

Table 8.1   Roles Required for Cost Management

Role	Responsibility
Customer lead	The member of customer's organization charged to manage the project from the customer's perspective. Has final authority over all additions, changes, or deletions to project scope, schedule, and performance.
Project manager	The person accountable for project success, the PM leads, monitors, and supports the cost management processes and procedures.
Procurement officer	Member of the project manager's team that can authorize new agreements and changes to agreements between the project team and the customer or between the project team and its vendors.
Subject matter experts	Members of the PM's organization who are engineers, scientist, and analysts with expertise in one or more area of the project.
Resource provider	Members of the PM's organization that develop and supply technical and project resources to a project team.
Control account manager	The person assigned as the single point of contact for one or more work element within the project. These work elements can be work packages, groups of work packages, or groups of human resources. The CAM may be a member of the PM's organization or the vendor organization if the work package is outsourced.
Financial analyst	A member of the project manager's team charged with the development and tracking of the project budget. Financial analysts typically have access to the financial management systems and can collect data required to generate project budget and expenditure reports. The financial analyst is educated and skilled in EVMS and is an EVMS SME on the project team.
Project scheduler	Member of the project manager's team that performs the consolidation, structuring, publication, and management of the project schedule.

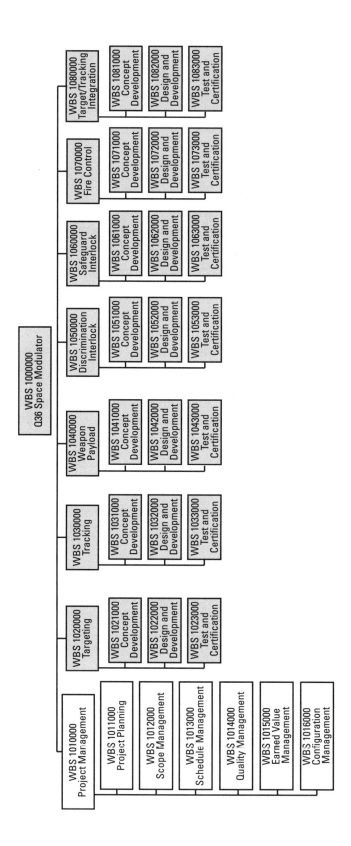

Figure 8.5 Example WBS.

Define project organization. The organization is depicted in a hierarchical structure detailing lines of responsibility through successive levels from senior managers to work package managers as shown in Figure 8.6. Within an EVMS, the organization structure contains the level of detail required to assign a single point of responsibility for each work package within the WBS and to identify unique rate structures for resources during the budgeting process. Resource providers (internal and vendor) will have rates for each skill set. For example, the rate for a systems engineer is different from that of a systems analyst and the rate of a schedule manager may differ from that of a quality manager. These rates will drive the cost incurred for each work package.

Organization charts are used in other processes such as communications and configuration management where additional performers and attributes can be added. This organization structure is used in the following steps in the assignment of work to responsible parties.

Integrate processes. The ability to link cost and schedule into a single set of performance criteria requires tight integration between each element of work (WBS element), the tasks associated with the work, the duration of work, the time frame the tasks are to be performed, and the cost to execute the work. All data points associated with cost and schedule must be linked to the same point of reference. This single point of reference is the project's WBS. As shown in Figure 8.7 the budget, schedule, resources, and tasks all align to the project's WBS. Note the columns preceding the budget columns were created during schedule management and are reused for EVMS-based cost management. Notice also that the resources provided in the assigned to column are consistent with the WBS in Figure 8.5.

Integrate WBS to organization. The objective of this step is to establish the linkage between each element of work and the single point of responsibility for the work. How these roles are assigned will vary and is typically driven by the

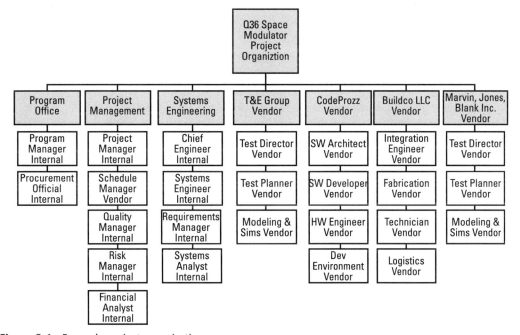

Figure 8.6 Example project organization.

	Work Breakdown Structure		Sequencing	Durations	Schedule		Resources		Budget Allocations			
WBS	Task Name	Predecessors	Successors	Duration (Planned)	Start (Planned)	Finish (Planned)	Resource Requirement	Budgeted Cost of Work	Contingency Reserve	Budget at Completion (BAC)	Management Reserve	Project Base Budget (PBB)
1000000	⊞ Q36 Space Modulator			2285d	07/02/18	04/02/27		$22,974,500.00	$2,297,450.00	$25,271,950.00	$2,297,450.00	$27,569,400.00
1010000	⊞ Project Management			2281d	07/02/18	03/29/22		$157,000.00				
1011000	⊞ Project Planning			95d	07/02/18	11/09/18	Program Manager	$22,000.00				
1012000	Scope Management	5		2275d	07/09/18	03/26/27	Project Manager	$8,000.00				
1013000	⊞ Schedule Management			2275d	07/09/18	03/26/27	Scheduler	$50,000.00				
1014000	Quality Management	9		1946d	10/14/19	03/29/27	Quality Lead	$24,000.00				
1015000	Earned Value Management	9		1946d	10/14/19	03/29/27	Financial Analyst	$12,000.00				
1016000	Configuration Management	9		1946d	10/14/19	03/29/27	MBJ LLC	$16,000.00				
1017000	Monitoring and Control	4		2275d	07/03/18	03/22/27	Project Manager	$25,000.00				
1011400	Work Authorizations	12	4	120d	04/12/21	09/24/21	Procurement Official	$13,000.00				
1020000	⊞ Targeting			410d	09/27/21	04/21/23		$2,180,000.00				
1021000	⊞ Concept Development			100d	09/27/21	02/11/22		$835,000.00				
1021100	Functional Requirements	18	22	20d	09/27/21	10/22/21	Requirements Manager	$100,000.00				
1021200	Architecture	21	23	20d	10/25/21	11/19/21	Systems Engineer	$120,000.00				
1021300	Systems Analysis	22	24	20d	11/22/21	12/17/21	Systems Analyst	$220,000.00				
1021400	Solutions Analysis	23	25	20d	12/20/21	01/14/22	Systems Analyst	$75,000.00				
1021500	Prototyping	24	27	20d	01/17/22	02/11/22	Turnkey System Prototyping	$320,000.00				
1022000	⊞ Design and Development			240d	02/14/22	01/13/23		$1,050,000.00				
1022100	System Requirements	25	28	60d	02/14/22	05/06/22	Requirements Manager	$100,000.00				
1022200	System Design	27	29	60d	05/09/22	07/29/22	Systems Engineer	$120,000.00				
1022300	Specifications	27	30	60d	05/09/22	07/29/22	Systems Engineer	$220,000.00				
1022400	Sub-systems	29	31,33	60d	08/01/22	10/21/22	Turnkey Design	$175,000.00				
1022500	Software	29	32	60d	08/01/22	10/21/22	SW Engineers	$320,000.00				
1022600	Integration	30, 31		60d	10/24/22	01/13/23	Integration Shop	$115,000.00				
1023000	⊞ Test and Certification			130d	10/24/22	04/21/23		$295,000.00				
1023100	Developmental Tests	29, 30		75d	10/24/22	02/03/23	Test Director	$75,000.00				
1023200	Technical Readiness	32		35d	01/16/23	03/03/23	Test Director	$100,000.00				
1023300	Operational Readiness	35		35d	03/06/23	04/21/23	Test Director	$120,000.00				

Figure 8.7   Sample project WBS integrated with schedule, cost, and resources.

organizational makeup of the individual resource group. In resource management it was discussed that resources will be assigned to the project in multiple methods based on how the resource provider operates and the type of work assignment. In some cases, a single person is provided by his resource manager and is designated as the single point of responsibility. When multiple resources are provided by a single provider, each may be assigned individual work elements to manage. In other cases, all work executed by an organization are managed by a single person within the department or vendor organization. In all cases, each work package or WBS element composed of several work packages will have a single point for cost, schedule, and performance management and reporting. This control point is referred to as a control account (CA) and the person charged as the single point of control is the control account manager (CA manager). Figure 8.8 is a graphic representation of this process. This format can quickly become too large to display and manage, so an alternative method will be provided in the next section.

Systems engineering in Figure 8.8 provides an example of how multiple work elements can be resourced by a single resource group that is managed by a single CA manager (systems engineering). In the other cases all work executed by the vendor or internal T&E group is managed by a single party. This nuance demonstrates the importance of establishing clear lines of responsibility for each element of work.

Develop Project Schedule and Budget

Schedule with network logic. Demonstrated in schedule management this is the process of identifying the sequence in which tasks must be performed, specifying dependencies between tasks, and estimating the durations of each task. The start

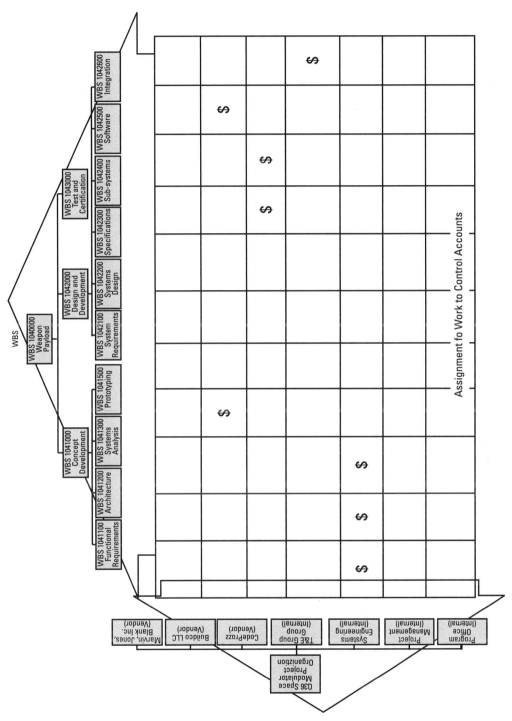

Figure 8.8 Control accounts: integration of WBS and ORG.

and end date for each task is then calculated based on these attributes, resulting in a logic-based project schedule. The network diagram and schedule created during schedule management are provided in Figure 8.9.

Establish budgets for authorized work. During the project initiation phase, cost estimates were created based on very high-level information about the project. At this point in the project planning phase, the PM now has a relatively mature understanding of the project scope (WBS with work packages) and the organization is growing to include more subject matter expertise and vendors. Each element of work has been assigned a control account manager charged with the execution of the work in Figure 8.10. A new cost estimate is created for each element of work using the expertise within the control account assigned to the work.

The cost plus, T&M, and FFP contracting methods were introduced in Chapter 7. How the vendors were contracted may drive the level of cost detail already available. In the cases where cost plus was used, the vendor provided a detailed cost basis with their proposals. Vendor proposals for FFP contracts will not include a high level of detail unless specifically requested by the contracting officer. FFP contracts are not appropriate for EVMS reporting. However, it is still wise to include these work elements in the EVMS as impacts and changes can still arise that will impact the rate at which value is earned. This concept will be further explained in the discussion about measurement indicators.

The next group of resources are those internal to the PM's organization. This group is the one that will require the most direct involvement of the PM as these resources are employed by the project with little-to-no risk acceptance by the provider. The arrangement with these internal organizations is like that of the cost-plus contract where the rates for each labor category are predefined and costs for management, travel, services, tools, and supplies are passed onto the project. The indirect cost associated with a given work package are included in the budgeted cost of work for the package. Therefore, it is very important to understand all labor categories and rates for those directly supporting the work along with the indirect costs for each work package.

The last cost element is that of the fees charged by the PM's organization for indirect costs, such as facilities, IT services, and administrative support. How these costs are charged to the project is typically defined by the PM's organization. Examples include a percentage charged against the project as part of the contract fixed fee, a single line item, or applying an apportioned amount to each WBS element. The key here is to ensure these costs are identified at this stage. Figure 8.10 is an example of a project budget with budgeted cost of work assigned to each WBS element, a total budget at completion (BAC) for the project, the portion held back for management reserve, and the project base budget (PBB) that reflects the total authorized budget for the project.

In Figure 8.10, notice also that a new column titled "CA Manager" has been added next to the resource column. This column represents the individual within the resource group that is the single point of contact (POC) for the work at this line item. By using additional columns in the project scheduling tool to support the budgeting and work assignments, all data is stored in a single data set. This level of integration up front during the planning stage enables a successful EVMS.

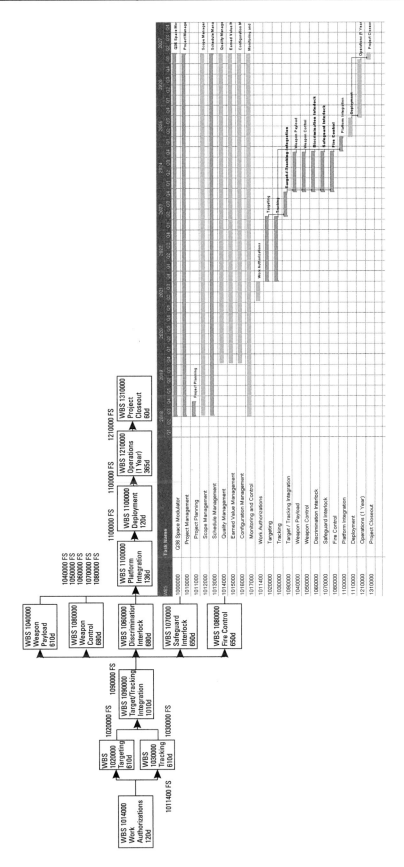

Figure 8.9 Sample network diagram and Gantt chart.

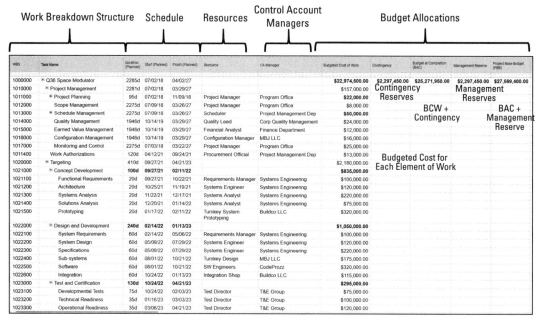

Figure 8.10 Project budget with BCW, BAC, and PPB.

A direct cost is that which can be directly traced to the work element or deliverable. It includes labor, services, and materials that can be traceable to the product. Labor is direct when it can be traceable to transforming materials or data into the finished product. Other direct costs include items in direct support of the work: travel, software, tools.

Indirect cost cannot reasonably be traced to the product and include utility bills, administrative expenses, insurance, and facilities or project management staff.

Sum individual control accounts. This step involves rolling up the individual costs assigned to each control account to establish a by resource view of the budget. By using a single tool for the WBS, schedule, resources, and budget lines, this step can be accomplished by creating additional formula columns. The BAC value and the total allocated budget will eventually be equal when all work is budgeted and a CA is assigned.

Define Measurement Methods and Baseline the Project

Measurement indicators are the methods used to calculate earned value during the execution of each element of work. Think of these indicators as the periods in which project snapshots are captured. As the nature of the work associated with each WBS element varies, the periods in which snapshots are informative also will vary.

The weighted milestones method breaks the WBS element into several submilestones to support earned value measurement. Each of these submilestones are assigned a portion of the WBS element's budgeted cost of work so that costs and progress data can be collected at increments according to the submilestones. Using the data in Figure 8.11, the following is an example.

The work package associated with WBS element 2.1.5 is contracted to Buildco LLC. Within the Buildco contract there are several milestones that divide the work package into billable portions of the work. This allows Buildco to recognize income during the project based on the completion of each milestone while providing the PM insight into Buildco's progress. A portion of WBS 2.2.5's BCWS is assigned to each of contract milestone. For this type of contract, the PM selects the weighted milestone method for measurement.

The fixed formula method recognizes earned value based on the start and completion of the WBS element. In this method, the task earns a percentage of the task's value when it starts and when it is complete. Example percent allocations are 0–100, 25–75, and 50–50 where the first number reflects the value for starting the task and the second number represents the value of completing the task. This method does not provide awareness of performance during the task, but may be applicable for those with a short duration [2]. Therefore, caution should be taken when selecting this method. Two examples are provided.

In a separate contract CodeProzz is contracted to deliver for WBS 2.2.4, which is a document that describes the subsystems within the targeting system. Due to the duration of this task the vendor's contract has a single payment milestone for delivery of the completed document. In this case the PM selects the 0–100 method for measurement.

A simplified acquisition of off-the-shelf equipment and materials (not shown in Figure 8.10) may be limited to placing the order and then accepting delivery of the products. However, the period between order and delivery is significant and the order must be placed very early to meet the schedule dependencies. In this case the PM selects the 25–75 method recognize the significance of starting the task.

The percent complete method uses the progress reporting provided by the CA manager to determine the earned value for the WBS element. According to a predefined reporting schedule, the CA manager reports the percentage of the work

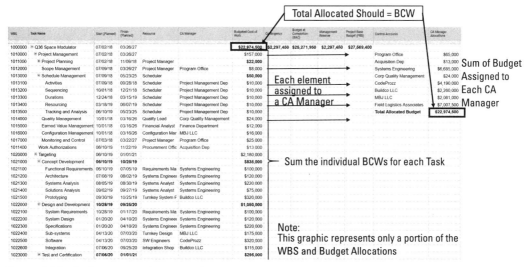

Figure 8.11 Budget summed to control accounts.

completed to date, which is used to determine the percentage of the budget that is recorded as earned value. For example, the budgeted cost of a work package is $100k and the CA manager reports at the first period that 25% of the work is complete. Therefore, the work package has earned 25% of its planned value or $25K. This method is often used in cases where it is difficult to break the work down into discrete milestones for the weighted milestone method or the milestones are separated by large periods of time and there is a desire to track progress between milestones. When using the percent complete method, the ability to detect and evaluate work completion is imperative.

The level of effort method can be used for planned expenditure rates of indirect costs to determine the earned value. Indirect costs are those that cannot be directly traced to a project outcome. In this method, each period is assigned a portion of the WBS element's total planned cost and at the end of each period that portion is recorded as earned value. Using the level of effort method, the BCWS and the BCWP are equal. This method is counter intuitive to the objective of EVMS as it does not demonstrate how the cost supports the project or its value to the customer. Therefore, this method should be limited to cases where no other method is achievable.

The apportioned value method provides the means for the PM to allocate a portion of project's indirect costs to each WBS element. This provides a means to demonstrate how indirect costs are related to the direct achievement of work. With this method, the indirect costs earn value based on their apportioned value to each WBS element as it earns value.

Project leadership. Project managers are often tempted to designate the cost associated with project management, the program office, or other leadership activities as indirect costs and then use the level of effort method to track and report these costs. However, customers are growing sensitive to costs they perceive as overhead within the PM's organization that they believe should come out of the contractor's profit or fixed fees. Additionally, this approach disincentivizes these members of the team from thinking about how their daily work relates to the value delivered to the customer. The PM should strive to identify the direct outcomes supported by as many of the project team members as possible. For example, within the project management section of the example WBS several of the activities performed by the PM team are listed as discrete efforts each assigned costs and schedule values. In the sample organization chart provided in Figure 8.6, the chief engineer is listed as a member of the organization. This person may not work directly on the production of deliverables, but he is expected to advise the SE team and review their work. These activities relate directly to the work performed by resources within his organization and should be included in the CA manager's estimate for that work package. By aligning what could be perceived as indirect efforts to the value of direct tasks, the PM can demonstrate how the costs associated with these resources relate to the overall project success and their value to the customer.

Establish performance measurement baseline is the last step in the planning and budgeting process. In Chapter 3, the project schedule was saved as a baseline that became the benchmark for measuring schedule performance. The same holds true with the EVMS version of the schedule and its allocated budget. This step locks down the budgeted cost (BCWS) of all work so that it can be used along with the budgeted cost of work performed (BCWP) and the actual costs (ACWP) to determine the projects performance indicators. This analysis will be demonstrated in the following sections.

Analyze Project Performance

Analyzing project performance involves the collection of performance data, calculating performance metrics, interpreting performance trends, and communicating performance to the project stakeholders. The key terms associated with performance analysis are provided in Table 8.2 to aid in this discussion.

Table 8.2 EVM Analysis Values

Acronym	Term	Formula	What it Means
ACWP (AC)	Actual cost of work performed.	The reported expenditures for the work that was completed. (Also referred to as AC actual cost.)	How much was spent on the work?
BAC	Budget at completion	Sum of the budgeted cost of work for all line items.	The total allocated budget for the project. (Does not include management reserves).
BCWP (EV)	Budgeted cost of work performed	BCWP = BCWS for completed portion of the work. (Also referred to as EV earned value).	How much work was completed?
BCWS (PV)	Budgeted cost of work scheduled	BCWS = The value of the work schedule to be completed up to a point in time. (Also referred to as PV planned value.)	How much work should have been completed?
CPI	Cost performance index	CPI=BCWP/ACWP	< 1.0: work is over budget for work performed =1.0: work currently on budget for work performed > 1.0: work is currently under budget for work performed
CV	Cost variance	CV= BCWP-ACWP	< 0.0: work is currently over budget = 0.0: work is currently on budget > 0.0: work is currently under budget
EAC*	Estimated cost at completion. Based on current trends	EAC=ACWP+((BAC-BCWP)/ CPI), or BAC/CPI.*	EAC > BAC project is over budget EAC = BAC project is on budget EAC < BAC project is under budget
SPI	Schedule performance index	SPI = BCWP/BCWS	< 1.0: work is behind schedule = 1.0 work is on schedule > 1.0 work is ahead of schedule
SV	Schedule variance	SV = BCWP-BCWS	< 0.0: work is behind schedule = 0.0 work is on schedule > 0.0 work is ahead of schedule
ETC*	Estimate to complete	ETC = BAC-BCWP*	How much will it cost to complete the project?

*Refer to the PMI practice standard for earned value management for addition approaches to calculating these values [2].

Collect performance data involves identifying the work elements that are in progress at a given period in the project schedule and recording the work completion and expenditures data for those items. As EVM involves the time-phased reporting and analysis of project performance, it requires snapshots of the data at intervals across the duration of the project. The duration between these snapshots can be monthly or quarterly, but anything more frequently can be labor-intensive and anything less frequently reduces the ability to respond to negative trends. For each reporting period, the scheduler and financial analyst work together to identify each WBS element that has started prior to the reporting date and based on the its defined measurement criteria designates the items in which data is required. The items deemed relevant to each reporting period are referred to as reportable elements. The budgeted cost of work scheduled (BCWS) is the percentage of the planned value that is scheduled to be completed prior to the reporting period.

Calculating performance indicators utilizes the data collected for each reportable WBS element to determine how the project is progressing against the planned schedule and budget. For each WBS element the following are calculated:

- *Cost performance index:* CPI=BCWP/ACWP
- *Cost variance:* CV=BCWP-ACWP
- *Schedule performance index:* SPI=BCWP/BCWS
- *Schedule variance:* SV=BCWP-BCWS

These processes are demonstrated using the sample project and the following stage setting data points:

- The project started its initiation and planning phases in July of 2018;
- The planning process resulted in the performance management baseline depicted in Figure 8.9;
- The vendor solicitation and contract awards are completed and direct work on the project deliverables started in July of 2019;
- It is now the end of September 2019 and the first relevant reporting cycle is upon the project team;
- The percent complete method is used for measurement of all WBS elements in this example except WBS 1.6 Monitoring and control, which is measured using the level of effort method.

The scheduler and the financial analyst identify the reportable WBS elements for this time using September 20, 2019 as the reporting date. This date is used to

Table 8.3   Example BCWS Calculations

Reporting Date	WBS #	Start (Planned)	Duration (Planned)	Days Elapsed	% of Duration (Planned)	Planned Value	BCWS
9/30/2019	2.1.1	6/24/2019	82	99	100%	$100,000	$100,000
	2.1.2	9/16/2019	82	15	18.3%	$120,000	$21,951

identify all tasks that should have started prior to this point. On other projects, this list of tasks may be further filtered to remove tasks that may have started but are under the fixed formula method and are not yet due any earned value. The final list of reportable elements includes 22 WBS elements of which some should be complete at this time and the rest are scheduled to earn a portion of their value at this point.

There are varying opinions as to how much time should elapse on a project before EVM metrics are of value as a decision aid. However, the collection of data should commence as soon as direct work commences. By collecting data and creating metrics early, the PM can validate the processes and data created by her EVMS.

The BCWS is calculated by determining the percentage of the duration that has elapsed between the task start date and the reporting date. If the elapsed time is greater than the planned duration, the BCWS is 100% of the planned value. If the elapsed time is less than the planned duration, the BCWS is the percentage of the planned value equal to the percentage of the planned duration that has elapsed.

Working with the CA managers, the scheduler and the financial analyst collect the BCWP and the ACWP.

These values are used to calculate the cost and schedule variances and performance indexes for each reportable element as well as for the overall list. This extra step of calculating the indicators for each line item is not often mentioned within other guides as only the overall values are charted and reported. However, adding these calculations to the PM tool will be very useful when identifying the root cause of poor results later in the process. Figure 8.12 depicts the results for some of the reportable elements and the overall project values.

WBS	Task Name	Duration (Planned)	Start (Planned)	Finish (Planned)	Budgeted Cost of Work	% Duration Elapsed	BCWS Current Period	% Complete	BCWP Current Period	ACWP Current Period	SV Current Period	SPI Current Period	CV Current Period	CPI Current Period
1000000	Q36 Space Modulator	2285d	07/02/18	04/02/27	$22,974,500	14.3%	$1,081,457.68		$869,791.84	$1,130,729.39	-$211,665.84	0.80	-$260,937.55	0.77
1010000	Project Ma... [1: Milestone: Project Start]			...6/27	$157,000	14.3%	$42,...		...	...		0.91	-$11,613.70	0.77
1011000	Project P...			...9/18	$22,000	0.8%	$22,... [5: Completed on time but over budget]					1.00	-$6,600.00	0.77
1011100	Project Initiation	0	07/02/18	07/02/18		100.0%	$0.00	100%	$0.00	$0.00	$0.00	0.00	$0.00	0.00
1011200	Project Charter	5d	07/02/18	07/06/18	$5,000	100.0%	$5,000.00	100%	$5,000.00	$6,500.00	$0.00	1.00	-$1,500.00	0.77
1011300	Planning Documents	90d	07/09/18	11/09/18	$17,000	100.0%	$17,000.00	100%	$17,000.00	$22,100.00	$0.00	1.00	-$5,100.00	0.77
1012000	Scope Management	2275d	07/09/18	03/26/27	$8,000	14.1%	$1,128.50	14%	$1,128.50	$1,467.04	$0.00	1.00	-$338.55	0.77
1013000	Schedule Management	1795d	07/09/18	05/23/25	$50,000	17.9%	$8,940.66	14%	$7,152.53	$9,298.29	-$1,788.13	0.80	-$2,145.76	0.77
1013100	Activities	60d	07/09/18	09/28/18	$10,000	100.0%	$10,000.00	100%	$10,000.00	$13,000.00	$0.00	1.00	-$3,000.00	0.77
1013200	Sequencing	60d	10/01/18	12/21/18	$10,000	100.0%	$10,000.00	100%	$10,000.00	$13,000.00	$0.00	1.00	-$3,000.00	0.77
1013300	Durations	60d	12/24/18	03/15/19	$10,000	100.0%	$10,000.00	100%	$10,000.00	$13,000.00	$0.00	1.00	-$3,000.00	0.77
1013400	Resourcing	60d	03/18/19	06/07/19	$10,000	100.0%	$10,000.00	100%	$10,000.00	$13,000.00	$0.00	1.00	-$3,000.00	0.77
1013500	Tracking and Analysis	1555d	06/10/19	05/23/25	$10,000	5.2%	$519.54	5%	$519.54	$675.40	$0.00	1.00	-$155.86	0.77
1014000	Quality Management	1946d	10/01/18	03/16/26	$24,000	13.4%	$3,215.86	11%	$2,572.69	$3,344.49	-$643.17	0.80	-$771.81	0.77
1015000	Earned Value Management	1946d	10/01/18	03/16/26	$12,000	13.4%	$1,607.93	11%	$1,286.34	$1,672.25	-$321.59	0.80	-$385.90	0.77
1016000	Configur... [3: Work Authorizations]			...6/26	$16,000	13.4%	$2,143.91	11%	$1,715.12	$2,229.66	-$428.78	0.80	-$514.54	0.77
1017000	Monitorin...			...2/27	$25,000	14.3%	$3,571.43	11%	$2,857.14	$3,714.29	-$714.29	0.80	-$857.14	0.77
1011400	Work Authorizations	120d	06/10/19	11/22/19	$13,000	68.1%	$8,849.40	54%	$7,079.52	$9,203.37	-$1,769.88	0.80	-$2,123.86	0.77
1020000	Targeting	410d	06/10/19	01/01/21	$2,180,000	19.8%	$515,000.00	[6: Behind schedule and over budget]					123,600.00	0.77
1021000	Concept Development	100d	06/10/19	10/25/19	$835,000	81.9%	$515,000.00						123,600.00	0.77
1021100	Functional Requirements	20d	06/10/19	07/05/19	$100,000	100.0%	$100,000.00	80%	$80,000.00	$104,000.00	-$20,000.00	0.80	-$24,000.00	0.77
1021200	Architecture	20d	07/08/19	08/02/19	$120,000	100.0%	$120,000.00	80%	$96,000.00	$124,800.00	-$24,000.00	0.80	-$28,800.00	0.77
1021300	Systems Analysis	20d	08/05/19	08/30/19	$220,000	100.0%	$220,000.00	80%	$176,000.00	$228,800.00	-$44,000.00	0.80	-$52,800.00	0.77
1021400	Solutions Analysis	20d	09/02/19	09/27/19	$75,000	100.0%	$75,000.00	80%	$60,000.00	$78,000.00	-$15,000.00	0.80	-$18,000.00	0.77
1021500	Prototyping	20d	09/30/19	10/25/19	$320,000	0.0%	$0.00	0%	$0.00	$0.00	$0.00	0.00	$0.00	0.00
1022000	Design and Development	240d	10/28/19	09/25/20	$1,050,000	0.0%	$0.00	0%	$0.00	$0.00	$0.00	0.00	$0.00	0.00
1022100	System Requirements	60d	10/28/19	01/17/20	$100,000	0.0%	$0.00	0%	$0.00	$0.00	$0.00	0.00	$0.00	0.00
1022200	System Design	60d	01/20/20	04/10/20	$120,000	0.0%	$0.00	0%	$0.00	$0.00	$0.00	0.00	$0.00	0.00
1022300	Specifications	60d	01/20/20	04/10/20	$220,000	0.0%	$0.00	0%	$0.00	$0.00	$0.00	0.00	$0.00	0.00
1022400	Sub-systems	60d	04/13/20	07/03/20	$175,000	0.0%	$0.00	0%	$0.00	$0.00	$0.00	0.00	$0.00	0.00

Annotation callouts on figure: 8: Cost Variance & Performance Index; 7: Schedule Variance & Performance Index

Figure 8.12 Example of EVM data collection and calculations.

1. The project started with project initiation on July 2, 2018.
2. Most of the work performed in this period has been focused on the project planning and organization.
3. In June of 2019, work authorizations were issued by the procurement official that allowed direct work on the targeting subsystem to commence. The procurement official will continue to issue work authorization for future tasks over the next few months.
4. It is now after September 30, 2019 and the scheduler and financial analysist have compiled the first set of EVM data. While it will more time to capture enough data to create meaningful trend, these early data collection drills serve to validate the EVMS.
5. The initial project management tasks were completed according to the schedule; however, the actual cost to complete these tasks has been higher than planned (over budget).
6. The work on the targeting concept development has been underway for several months and several of the tasks should have been completed. However, the report shows the actual amount of work completed (BCWP) is less than planned and these tasks are behind schedule. Additionally, these tasks are also spending money faster than planned, as the ACWP is less than the BCWP for these tasks.
7. The top value for the schedule variance (SV) reflects that the project is behind schedule by $211,665.84 and the schedule performance index (SPI) is .80. Any SPI below 1 indicates the project is behind schedule.
8. The top value for the cost variance (CV) reflects that the project is over budget by $260,937.55 and the cost performance index (CPI) is .77. Any CPI below 1 indicates the project is over budget.
9. Based on current performance, the project will experience a budget shortfall of $260K even if all remaining work is on budget. The team will have to meet the budget on every future task and be $260K under budget on a mix of the tasks to finish on budget.

While these indicators are less than desirable, the project is just getting underway and new project teams do take time to get their processes refined before they start to work at their targeted performance. For this initial reporting cycle, the goal is to test the process more than to guide decisions. For this period the PM recommends to her leadership that these values should not be alarming and she will work with her CA managers to ensure they are receiving the support required to improve performance. The PM's leadership agrees but stresses the need to meet again at the next reporting period.

This process is repeated by the project team at the end of each 3-month period for the duration of the project.

Visualization and interpretation of the EVMS data as demonstrated in the previous section is good for one reporting period. But the value for using an EVMS is data is captured at recurring periods of the project enabling the visualization of trends for these periods.

Assume the sample project has now advanced another 18 months and the project team has collected and calculated the EVM indicators at each 3-month period.

The collection of historical data can now be graphed to help visualize the project's performance. Figure 8.13 reflects the project team's performance over the following 18 months.

Unfortunately, the team has not been able to overcome its challenges and continues to underperform in both cost and schedule. Figure 8.13 shows the planned value of work for each period (BCWS) on the center line that continues until the end of the project. The bottom line depicts the value of the work that was completed for each period up to this point in the project (BCWP), which is below the BCWS line, indicating less than the planned amount of work is being completed (earned). The top line depicts the actual costs incurred (ACWP) to achieve the work earned on the BCWP line. The fact that the ACWP line is above the BCWP line indicates the amount of money being spent is more than the amount of value being earned. At several points the ACWP orange line is above the BCWS. In this case it means that in addition to spending more than is being earned, the team is spending faster than was planned. The ACWP line reflects negative performance regardless of its relationship to the BCWS line whenever it is higher than the BCWP Line. Unfortunately, the PM was not able to correct the situation and one could expect her meetings with leadership in June and September of 2020 were very stressful.

What would it look like if things did not continue as shown in Figure 8.13 and the results were better? For example, if in the period leading up to 12/30/19 the PM was able to work with her CA managers to help get costs under control by reducing staff during wait periods caused by late completion of earlier tasks. Figure 8.14 shows that after 12/20/19 the project continues to achieve value later than expected but the cost expenditures (ACWP) are now trending below the BCWP. In this case, work is behind schedule, but costs are now being expended at a slower rate.

Unfortunately for the PM in this example, the customer did not like the wait and see approach recommended back in September 2019, and in October the PM's leadership was forced to bring in a more senior project manager in October of that year. Using her vast troubleshooting experience the new PM was able to identify the root cause of the team's performance and immediate results were demonstrated

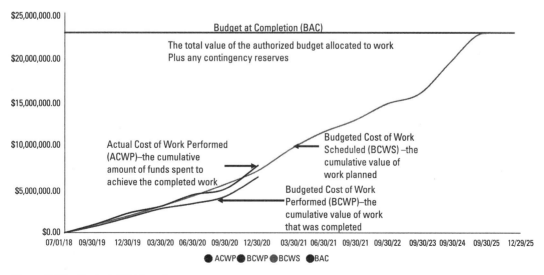

Figure 8.13 Project EVM trends.

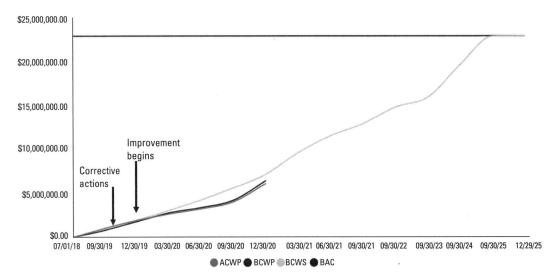

Figure 8.14   EVM trend: Behind schedule under budget.

in the following months. The new PM was able to eventually get the project ahead of schedule and on budget as shown in Figure 8.15.

These examples demonstrate how performance trends can be depicted and used to determine the need for corrective action or if actions taken are improving results.

Summary

Earned value management is an objective method of determining the health of the project based on integrated project data. The EVMS is the collective processes used to structure the project, collect data, and transform the data into actionable information. EVMS provides the PM insight on the current and past performance of the

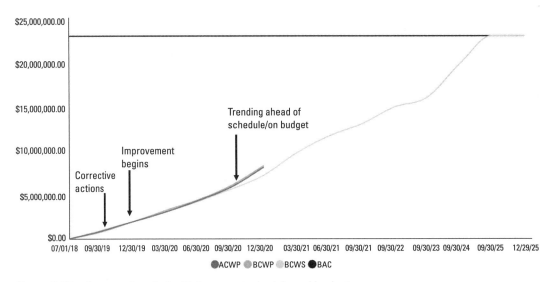

Figure 8.15   Recovered project with improved schedule and budget.

project in terms of budget and schedule based on the value earned for each WBS element. The PM can use the past performance indicators to estimate the future performance of the project and to determine the level of performance required to get the project back within its cost and schedule constraints.

EVMS provides the project leadership team and customer a level of awareness not available with conventional cost and schedule monitoring. EVMS is a systematic approach to the integration and measurement of cost, schedule, scope, and risk which can guide the successful achievement of project objectives.

By using the common sense best practices described within this book for all projects, the PM does not require a secondary planning and development process when EVMS is desired. The situational awareness and decision support provided by the PM will greatly increase the PM's value to her leadership and customers.

References

[1]   American National Standards Institute (ANSI) and Electronics Industry Association (EIA), *ANSI.EIA Standard 748c Earned Value Management Systems*, New York: American National Standards Institute, 2013.

[2]   Project Management Institute (PMI), *Practice Standard for Earned Value Management*, Newtown Square, PA: Project Management Institute, 2013.

[3]   Lillard, J. D., *Earned Value Management Systems: Example Models*, Charleston, SC: Capabiltiy Assurance Institute, 2019.

CHAPTER 9

Configuration Management

Throughout the project life cycle, items in the form of information, data, documents, prototypes, software code, and final deliverables are created that directly or indirectly lead to the satisfaction of the project's expected deliverables. Within these items are a set of items that require their configuration to be managed to ensure the most recent version of the item is available to the project team and that any changes to these items follow a structured, thought-driven approach.

Configuration management is the systematic process of maintaining a single set of these configuration managed documents, data, systems, and products associated with the project. The objective of configuration management is to ensure the project team

- Works to the currently approved scope and contractual conditions;
- Has a set of requirements that accurately reflects the approved scope;
- Designs project deliverables consistent with the approved requirements;
- Develops products and services that reflect the currently approved designs;
- Delivers products and services that conform with the project scope, requirements, and designs;
- Documents the configuration of the currently delivered products and services;
- Has available documentation of the current configuration to maintain and support fielded products.

A change to of any of these items will impact the configuration items (CIs) produced in each of the downstream processes of the project. Configuration management is the set of processes used by the project team to identify CIs then ensure that any changes to these CIs are managed, tracked, monitored, and communicated.

New technical project managers require an understanding of configuration management and the methods used to manage changes that affect the success of technical projects. The following resources are used in this chapter and are recommended reading.

The NASA Systems Engineering Handbook, Rev2. National Aeronautics and Space Administration. Washington DC, 2007. This publication distills the complex topics associated with systems engineering and space systems development into a structure understandable to nontechnical audiences. This document is free from the NASA.gov website [1].

The INCOSE Systems Engineering Handbook, A Guide for System Life Cycle Processes and Activities, Fourth Edition. International Council on Systems Engineering, San Diego [2].

Configuration management, like the other processes, occurs throughout the entire project life cycle from initial planning to project closeout. CM is also cyclical in its execution as even the plan for configuration management and its processes and techniques may require adjustment as the project progresses. Configuration management is integrated with the other project management and systems engineering processes used by the project team. Figure 9.1 depicts the common processes associated with configuration management as derived from the referenced publications.

This chapter introduces the new technical project manager to configuration management as a critical element in the successful delivery of products and services meeting the project scope and customer expectations. These processes require specialty knowledge, skills, and tools in which the PM relies on a CM practitioner to support the project. As the person accountable for the project's success, the PM plays an important role in the leadership and integration of this process.

The roles depicted in Figure 9.1 demonstrate the integration between configuration management and the other project management and systems engineering processes. The CM lead plays an integral role in these other processes and is an important member of the teams executing these processes. Effective project configuration management is achieved when the entire project team recognizes the importance of managing and communicating change within all project activities.

Plan Configuration Management

CM planning involves the documentation of standards, policies, practices, procedures, and instructions that guide the project team in managing the configuration

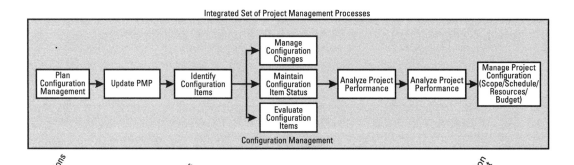

figure 9.1   Configuration management process.

of its final deliverables as well as the intermediate information, mechanisms, and assemblies that lead to final products. The CM lead looks to the organization's process assets to identify standard processes, tools, and templates to support the project. During the planning processes, business rules and thresholds are defined that drive the what, when, and how each item created and used by the team enters the CM cycle. During CM planning, the PM and CM lead identify the of types items that will require configuration management.

Contractual documents and data. The project's scope and contract documentation are the foundation for all work performed by the project team. Ensuring the project team works to the currently approved scope and contractual conditions requires these documents be included in the CM planning and management processes. In addition to describing the work associated with delivering the products and services, these documents include lists of supplemental reports, submissions, and data that must be supplied under the contract. The lists of supplemental information represent scope for the project team can be very extensive and costly to satisfy. The following are examples of the lists used in contracts.

- *CDRL*: A list of data, artifacts, and reports that must be submitted to the customer.
- *SDRL*: A version of the CDRL used between a contractor and its subcontractors.
- *DID*: A document that specifies each data element listed within the CDRL or SDRL. These documents define the data content, format, and intended use for each data element.

During CM planning, the PM and CM lead identify all data deliverables associated with the project and select the processes required to maintain CM on the contract documents and all contractual data deliverables.

Requirements data. Requirements management is change sensitive as the goal of the process is to refine and clarify customer needs into clear requirements for the development and production of products and services. Requirements reflect the currently approved scope and are used to develop the designs. Those involved in the identification and management of requirements data management need to know when a change or addition of requirements triggers the CM process. This integration between process areas helps to ensure the project team has a set of requirements that accurately reflects the approved scope and designs project deliverables consistent with the approved requirements.

Internal data and documents. The planning process defines the point in the development process in which planning documents enter configuration control. One may assume that only items delivered to the customer constitute configuration items. However, each discipline creates information and artifacts that drive the creation of final deliverables. These items go through multiple internal reviews and edits during creation and may require an internal CM process to maintain consistency and traceability. Careful attention to the impact any change to internal documents may have on downstream users of the documents will help the team ensure it is working from the same set of data.

Project review cycles. The use of periodic reviews was introduced in Chapter 3 to ensure stakeholder input is provided on plans, designs, and products. These reviews generate comments, questions, and concerns that could lead to changes in interim and final products. All input received during these reviews requires a process to facilitate the review, adjudication, and communication of these comments. The information collected during this process serves as a record of the review and ensures stakeholder input is acknowledged. In the event the stakeholder input necessitates a change, this information becomes an input to the change control process and establishes traceability within the product information. The CM plan includes the processes followed for recording internal and external product reviews and for identifying when these processes result in a potential change to the project scope.

Design evolution. The design and development of products and services evolves through a natural flow that starts with a customer need, which is translated into requirements that are then used to create a design for the development and production of the solution. The project life cycle does not end until the solution is accepted by the customer and in some cases, and it does not end until the solution has reached the end of its usefulness to the customer.

The term phases and stages are used to define each increment of the project life cycle. At each of these stages, the maturity of the data, designs, software, assemblies, and final products is represented as a set of CIs for that stage. The CIs created for each stage are saved off to a baseline that can then be referred to in later stages of the project. Team members working in subsequent stages of the life cycle use these baselines as the starting point for their work. Any modifications and changes to these configuration items become new CIs associated with that phase of the project. The names and composition of these baselines will vary from project to project based on the lexicon used by the design team and the project life cycle. It is very important to document the selected lexicon and maturity expectations associated with each baseline of the product's design, development, and delivery.

The NASA Systems Engineering Handbook defines four typical baselines: Functional, Allocated, Product and As-deployed. Each of these baselines has a specific meaning and set of expectation during the product life cycle [2].

Life cycle approach. Projects frequently adopt a product development life cycle in which customer solutions are delivered in increments, phases, or cycles. Each of these increments constitutes a subproject that runs through the project life cycle and then repeats for the next increment of capability. The configuration management plan addresses the management of CIs during each of the life cycles and how the project will differentiate between [2].

Configuration management organization. The processes and procedures adopted for each of the previously described configuration items are tailored to strike a balance between too little and too much regulation imposed on the project team. These processes define the parties involved in the review and approval of changes depending on the type of change and level impact on the project. Configuration control boards (CCBs) are the organization construct used to manage the configuration control process. Multiple CCBs can be established with each representing a level of change control authority. For example, the systems engineering team may have a local CCB whose authority is limited to adjudicating comments or refinements to the systems design that does not change the scope of the project while the enterprise level CCB must approve any element of the project design that does not comply with the customer's enterprise architecture, while any changes to the project scope, system requirements, and contract must be approved by the project level CCB.

Identify Configuration Items

Based on the processes and procedures defined in the CM plan, the project team designates items such as work products, designs, production lines, software code, or deliverables as configuration items. The CM lead helps the owner of each item through the process, but the process is typically executed by the item owner. Each CI is assigned a unique identifier (CI-ID) that is used to track the CI within the CM data set. The method of assigning CI-IDs can vary from a file-naming structure to system-generated alphanumeric codes. The CI-ID is placed on physical elements and inserted within the structure of documents so that the item can be identified and referenced.

Each CI is logged within a CM data set that includes attributes describing the functional or physical characteristics of the CI, the type of baseline the CI represents, its owner, and any proposed or approved changes to the CI. Depending on the physical characteristics, the document, software code, or data set is checked into a repository and protected from unauthorized changes. Team members can continue to work on future baselines using a copy of the CM managed item, but these changes are not official until the new CI runs through the CM process resulting in a revision to the existing CI or the designation of new CI.

During the CM planning process, the PM and CM lead identified contractual documents and data delivery requirements associated with the project. These items immediately enter the project CM cycle, are assigned CI-IDs and are logged within the CM dataset.

Manage Configuration Changes

Changes to configuration items can originate from many sources as shown in Figure 9.2. Seemingly innocent communications between the customer and the project team can lead to changes in a CI that left undocumented could impact the project. Managing any changes to CIs helps the project team to ensure that all proposed changes follow a deliberate process to understand the potential impacts on the project's cost, schedule, and performance constraints.

Once an item is designate as a CI under CM, any changes to the CI must follow a deliberate process to determine the need for a change, the impacts of the change on other CIs and implications of the change on the project. CI change management is not intended to stop all changes, rather change management is intended to minimize the consequences of change and to ensure the change is documented and communicated. Figure 9.3 depicts the process followed to control change during configuration management and its mutual dependence with the other PM processes.

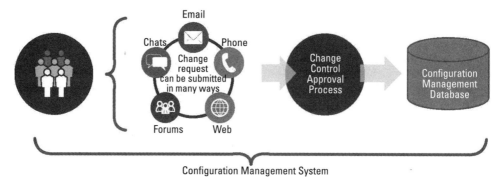

Figure 9.2   Sources of change conditions.

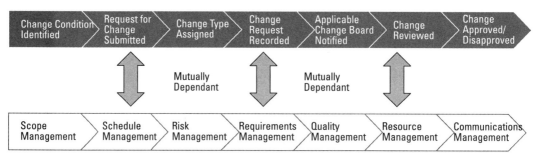

Figure 9.3   Change management process.

All changes to CIs are documented using the project's RFC process and templates. The complexity of the change and its impact on the project and the customer's organization determine the reviews and authorities required for each change. Therefore, it is important for the project team to have a clear set of ground rules for categorizing CI change proposals so that the proper reviews and authorities are applied to the proposal. These categories are defined during the planning process and may be added to or revised as new conditions arise. Table 9.1 provides an example of change categories and the authorities required to approve [1, 2]. Each project will define or adopt categories based on the customer and PM organizational standards or policies.

Technical projects often use a series of RFC templates common to the engineering community. The ECP is used to reflect changes to the product requirements, design, or the form and function of the deliverable. These changes would fall into the Class I and II types in Table 9.2. These changes may impact the project's cost and schedule positively or negatively. For example, incorporating technology or material different from those specified may reduce cost and complexity of the solution while achieving the same performance. Even with perceived positive impacts the change must still be documented, reviewed, and approved before the engineering team can implement the change.

Configuration changes that can be approved locally such as a Class III type in Table 9.2 still require notification to those impacted by the change. ENs provide the format and process to ensure minor changes are communicated to the consumers of the affected CI. For example, the structural engineer has corrected a dimension noted on one of his drawings. The change follows the Class III approval process

Table 9.1 Configuration Management Roles

Role	Responsibility
Customer lead	The member of customer's organization charged to manage the project from the customer's perspective. Has final authority over all additions, changes, or deletions to project scope, schedule, and performance.
Project manager	The person accountable for project success, the PM leads, monitors, and supports the project's configuration management processes and procedures.
Procurement officer	Member of the project manager's team that can authorize new agreements and changes to agreements between the project team and the customer or between the project team and its vendors. The PM and CM lead work with the procurement officer to ensure CM requirements are appropriately included vendor solicitations and contracts. The procurement officer also ensures all changes that impact the current agreements are executed according to the organization's policies.
Configuration management lead	Member of the project organization that leads the development and execution of configuration management plans, processes, and procedures at the project level.
Project quality lead	Member of the project organization that leads the development and execution of quality management plans, processes, and procedures at the project level. The quality lead plays an important role in the CM processes.
Engineers/analyst	Members of the project manager's technical team that collect, elaborate, express, and analyze proposed changes to CIs. These technical members of the project team review each change request to identify all points of impact and to determine the impact of the change on the technical plans and the system.
Requirements manager	Member of the project manager's team that performs the consolidation, structuring, publication, and management of requirements. Requirements management is change intensive as requirements are continually elaborated and refined throughout the project life cycle. Not all changes to requirements constitute a change in scope but the currently approved requirements data set is a CI. The CM lead works with the requirements manager to establish thresholds for when changes to requirements must follow the project's CM processes.
Risk manager	Member of the project manager's team with training and experience in risk management. The risk manager provides guidance in the identification and analysis of project risks then manages the consolidation, structuring, publication, and management of the project risk plans and data. Configuration management reduces technical risks by ensuring the project team is working to the current project definition and producing products that conform to this definition.
Configuration item owner	The member of the project team that servers as the single point of contact for a given configuration item. This person is someone directly involved in the item's creation or development. For example, the requirements manager may be the CI owner for the requirements database and the risk manager for the risk registry.
Configuration control board(s)	One or more organizations within the project team charged with the review, analysis, and approval of requested changes to configuration items.

and the EN is used to notify all consumers of the drawing that a new version is now available in the document repository [2].

All configuration change requests are reviewed to determine first if the change proposed is required and then for the impact this change will have on other CIs. The parties involved in the configuration change proposal are first dictated by the class of change, which defines the CCB for the change. Based on the initial review by the CCB, other parties may be added to the review to ensure all potential impacts are identified. The results of the configuration change review are then reported back to the CCB, which then approves the change, requests further analysis, or rejects the change request. At each step in the process the status of the change request is logged in the CM data set.

Table 9.2 Example Change Classifications

Common Name	Description
Class I	A change that is significant and may impact on the customer's organization, enterprise architecture, or safety. These changes require approval by the customer. Configuration deviations or change requests to the customer are most likely approved at some sort of formal or informal CCB at this level in which the project PM is typically not a participant.
Class II	A change that affects the cost, schedule, or performance of the project, PM's organization or the customer's organization. These changes typically require approval by a CCB that includes the PM and his or her leadership team and SMEs. There may be cases where this level CCB must defer to the customer for approval.
Class III	A minor change that reflects corrections to documents or designs. These changes are typically approved by the CI owner but are still documented according to the project change control process. Some CI owners establish local CCBs to help facilitate the review and awareness of proposed changes. The product reviews introduced in the previous section often fall into this category.

From: [1, 2].

There is a special class of configuration change request that applies to conditions in which the specification, terms, or conditions cannot be met by the project team. These conditions are explained in more detail in Chapters 7 and 10 where the vendor or the project manager requests a waiver for a product that is not compliant with customer's expectations. This condition may be permanent or temporary but still constitutes a need to document a configuration change condition for the project. Project deviations are a form of configuration change condition used when authority to deviate from the currently approved configuration cannot be met temporarily. Approval of deviations is temporary and has an expected date in which conformance with the approved configuration can be achieved. Waivers are used in cases where the project team has determined that a requirement cannot be met within current cost and schedule constraints. Waivers grant approval to the project team to remove or lower the requirement while still providing a usable solution. Following the configuration maintenance process ensures these requests are documented, reviewed, and approved.

Maintain Configuration Item Status

Throughout the CM cycle there requires an accounting of each CI and its status. This process ensures approved changes are carried out and the results are documented. Each CI is tracked in the CM dataset which is viewable by all members of the project team.

Each new request for change is logged against the CI along with the results of each step in the change management process. The notes, forms, and meeting minutes for the change process are also collected and stored in the CM repository.

Changes approved by a CCB are executed against the CI associated with the change request. These changes are then carried through to each document, design, SW code set, or any other affected CI. These changes can take time to implement across all the affected CIs. The actions assigned to each CI owner must be tracked to ensure all changes are accounted for in the CM data set. The status attributes for each CI are updated and the latest version of the CI is loaded into the CM repository before any change is considered complete.

Note: Many publications refer to this step as "configuration accounting" or "configuration audits." This chapter uses Maintain Configuration Item Status as it better reflects the purpose of this process.

Evaluate Configuration Items and Changes

Effective configuration management requires continual monitoring and evaluation of the CIs to verify that the CI has not encountered unauthorized changes and that all approved changes have been accurately implemented. These evaluations always occur whenever an approved change has been reported as completed by the CI owner.

Project reviews present additional opportunities to evaluate configuration items. Evaluations prior to each review can help the PM to verify that each product slated for review meets the criteria expected for the product at that point in the project life cycle. This is also an opportunity to ensure that all products slated for the review have been established as configuration items within the configuration management data set. Lastly, all CIs preparing for final delivery to the customer require an evaluation to verify that the delivered configuration matches that of the controlled configuration. This step helps to ensure the team delivers products and services that conform with the project scope, requirements, and designs.

Release New Configuration Items

Releasing a new system version, capability, or services absent of a structured release management process can have the same negative impacts on project success as uncontrolled and undocumented change. Rarely is a new or modified version of a CI delivered to the customer independently of other items. Often a configuration item is delivered or put into service as part of a formal rollout or incremental delivery. For example, a new version of a software package is ready for deployment at the customer locations. This new version requires updated user manuals, revised help desk procedures, and additional training for the new features to be effectively utilized by the customer. A sudden unannounced change in the software configuration will result in increased calls to the help desk as the software no longer works as expected. Without a coordinated release, the help desk is not aware that the software has changed and the version of software at the help desk may not match that of the users. The result of this uncoordinated rollout is a high level of customer dissatisfaction, a help desk that is incapable of helping the users, and the perception that the product has failed. These releases must be planned and coordinated across all parties delivering CIs associated with the new version of the software. Careful attention to the coordinated release of products to the customer helps to ensure that all documentation reflects the current configuration of the delivered products and services, which in turn improves the team's ability to maintain and support fielded products.

Summary

To achieve configuration management objectives, the project manager must ensure CM is integrated into the project life cycle and the other processes executed by the project team. The PM requires the support of a person trained and equipped for the planning, implementation, and monitoring of CM for the project. This support is provided by a CM lead assigned to the project team.

The PM and his CM lead develop the CM planning documents that define when an item should enter the CM cycle as a CI and how the CI will be documented, tracked, and managed within the CM cycle. All contractual and scope documents are CIs managed by the CM processes. The CM plan defines the types of changes that may occur in the project, what constitutes a change condition, and the authorities required to review proposed changes

In the event of a change condition, the change management process is used to track the change condition and ensure the proper parties have reviewed the change before it is approved. The approval and review process for each change condition is managed by a CCB with the authority required for the subject change. As all changes have the potential to impact the cost, schedule, and performance of the project, the CM process is interdependent with the other project management processes.

Change is inevitable in any project, but the impact of change on the project can be managed. The PM minimizes the negative impacts by establishing and leading a robust set of CM processes, procedures, and controls for managing change.

References

[1]   National Aeronautics and Space Administration, *NASA Systems Engineering Guide*, Washington, DC: National Aeronautics and Space Administration, 2007.

[2]   International Council on Systems Engineering, *Systems Engineering Handbook: A Guide for System Life Cycle Processes and Activities*, Hoboken, NJ: John Wiley & Sons, 2015.

CHAPTER 10

Quality Management

A common theme throughout this publication is that project management leads to the delivery of a product or service that meets the customer's requirements. Requirements define what the product or service is to perform and how well the product or service should perform. In Chapter 4, mature requirements are described as complete, singular, clear, necessary, and measurable. Quality management is the set of processes that help to enable the successful delivery of products and services that satisfy the customer's requirements. The objective of quality management is the satisfaction of customer requirements in a consistent, efficient, and effective manner. Multiple books, processes, frameworks, and standards have been published on the topic of quality. As an introductory chapter on quality management the following sources are provided:

> *ISO 9001:2015 Quality Management Systems–Requirements.* This standard defines the requirements for a quality management system and is used as the basis to validate an organization's quality management system [1].
>
> *Quality Management Principles.* This free publication available from the ISO website describes the seven quality management principles (QMPs) on which the organization bases its ISO 9000 publications [2].
>
> *ISO 9001:2015 PowerPoint Presentation.* This 15-page presentation provides the reader with an introduction to ISO 9001 and how the latest standard can be applied to organizations of all sizes. This document is also free from the ISO website and may help the PM to understand how his organization's quality program is intended to help individual project success [3].

Most publications, including those provided above, define quality management as an organizational endeavor that establishes quality management as an enabler to the organization's success. According to the ISO 9001:2015 Standard, a quality management system (QMS) includes quality planning, quality improvement, a set of quality policies, and objectives that will act as guidelines within an organization [1]. This nuance is important because project management publications, including the *PMBOK Guide,* assume the existence of a QMS available to the project manager. The assumption is that the PM will have an existing set of processes, tools, and expertise that can be tailored for the specific project [4]. These organizational process assets associated with quality management may or may not be based on a QMS as defined by the ISO 9001:2015 Standard. This chapter provides the new technical project manager with an introduction to quality management as a tool in the successful delivery of products and services to customers. This chapter also assumes that the PM's organizational process assets include quality management, which may or may not include a formal QMS as defined by ISO. The term QMS is used here as a generic term for an organization's quality processes, tools, training, and practitioners available to the PM.

Quality management involves the activities associated with planning for quality, assuring quality, and controlling quality, as shown in Figure 10.1.

Integration of quality management into any project provides multiple benefits that should not be ignored. These benefits include the reduction of waste in processes, improvement of product and service quality, faster turn-around time (TAT), lower costs, and increased customer satisfaction.

Quality management occurs throughout the entire project life cycle starting with understanding customer requirements during project initiation and ends only after the successful achievement of the project's objectives.

Often, the terms quality assurance and quality control are incorrectly interchanged in conversations about project and quality management. Quality assurance is the set of proactive activities aimed at ensuring the processes and mechanism used in the creation of products and services minimize the potential of defects and rework. Quality control is the activity that manages quality by inspecting the quality of a product or service after completion. Quality control used without quality assurance results in quality achieved only through rework, customer rejects, and the loss of time and money. However, used together these two methods can reduce the potential for nonconforming products and services while having the safeguards in place to quickly detect and resolve errors.

Quality assurance prevents defects caused by the process used to make the product (preventive process)

Quality control detects defects in the finished product (reactive process)

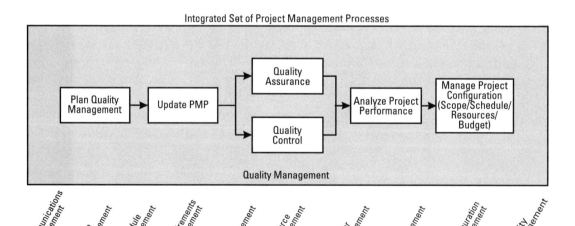

Figure 10.1 Quality management processes.

In Chapter 7, quality methods were used to describe the people, processes, and tools used to verify products purchased from a vendor meet the contractual requirements. Quality management is the set of processes that resulted in those methods. However, quality management is used by the project team for all products and services delivered by the team internally and externally.

Also mentioned in Chapter 7 was the potential for a supplier to be ISO 9001 certified. A company that holds a certification is one that has had its QMS assessed by an independent verification agent and has been deemed to meet the requirements in 9001:2015. While not limited to manufacturing, the ISO 9001 certification is frequently included in contracts that involve the manufacture of products, assemblies, or systems for a project. Using the ISO certification as a proposal evaluation criterion in these cases gives the buyer a higher level of confidence in the vendor's ability to deliver quality products in an effective and efficient manner. To the project manager this could equate to a reduction of risk. Therefore, effective quality management also involves integration with the risk management activities defined in Chapter 5.

Quality management is a specialty knowledge area and the PM should seek a qualified quality practitioner to lead the planning and implementation of quality processes for the project team. An organization with a mature QMS will have these experts available to the project manager. The project team includes a quality lead that ensures the project's quality plan is in place and supports the project by working with the entire project team to deliver products and services that meet customer requirements.

Quality management plays an important supporting role in every project. The PM works with the quality lead and his organization's quality management department to ensure a quality management system is in place early in the project life cycle. The QMS must be in place in time to support the planning and design of the production processes and mechanisms to ensure quality is a component of the designs.

Quality planning involves the documentation of standards, policies, practices, procedures, and instructions that will act as guidelines for the project team to deliver quality products and services. The typical outputs of quality planning include the quality management plan, which defines the processes followed for performing audits, tracking quality performance, performing product inspections, and the use of corrective action plans along with the roles and responsibilities for each of these areas. The planning also defines timing for quality audits, training for the project team, and the tools utilized for quality performance tracking and measurement. Mature quality management organizations have an organizational quality management plan and project quality management templates that can be tailored for the project.

Quality assurance (QA) prevents defects from occurring during the production of products and services. QA is the proactive set of activities within a balanced project quality management system. Quality assurance strives to include quality as a component of the definition, design, production, and testing of the project deliverables. Effective quality assurance can help to mitigate the risk of product failure and rejection. Product failures and rejections negatively impact the project's budget and schedule due to the time and money required for corrective actions. It is for this reason that the quality lead plays a significant role in the project's risk management team. Quality assurance identifies the things that could go wrong in the production cycle and then implements measures aimed at preventing these things from occurring.

Quality assurance also includes the continual assessment of the organization's conformance to the defined processes and procedures. Audits of production process and its performers serve to verify these processes are being followed and that the process still supports the intended purpose.

Quality audits also serve to identify inappropriately modified processes. Frequently, changes in the production environment, raw materials, resources, and other conditions impact the ability to execute a process or the process as designed fails to meet the intended results. The undesirable result is the performers self-adjust their processes to meet the new conditions without using process improvement techniques. When not approached in a deliberate method that includes quality as a consideration, the new process may have unintentional negative impacts. That is if a new process is defined. In many cases the individual performers will create their individual adaptations to the process resulting in inconsistency in the end results. Neither of these conditions lend to the consistent delivery of quality products and services. Therefore, quality assurance also includes process improvement practices used to identify the root cause of the process failure and the modifications to the process that will enable effective, efficient, and consistent outcomes.

The ability to define effective and efficient processes, identify root causes, and to objectively audit conformance to processes requires special skills and a certain level of independence from the organizations performing the processes. A PM must ensure that his quality lead is supported by his organization's QMS and has the autonomy to report observations without fear of retribution or impediment from the project team. Data resulting from this process includes performance metrics, audit reports, process improvement recommendations, and process improvement results.

Quality control (QC) involves the validation that project deliverables conform to the customer's requirements. QC is less proactive than QA but is does play a role in the identification of needed improvements. Quality control uses inspection and testing to determine product conformance prior to delivery to the customer. It also tracks failures and rejections after customer receipt to identify potential indicators of product or service quality issues. Data resulting from this process area includes discrepancy reports, corrective action reports, and nonconforming material or process reports.

The roles depicted in Table 10.1 demonstrate the integration between quality management and the other project management processes. The quality lead plays an integral role in these other processes and is an important member of the teams executing these processes. Effective project quality management is achieved when quality is an integrated practice for the entire project team and all activities within the project.

Plan Quality Management

Quality planning involves the activities required to establish quality as an integrated practice within the project team. Quality planning identifies requirements that define the quality of the project deliverables and then describes the quality

Table 10.1   Roles Associated with Quality Management

Role	Responsibility
Customer lead	The member of customer's organization charged to manage the project from the customer's perspective. Has final authority over all additions, changes, or deletions to project scope, schedule, and performance. The customer lead also serves as the arbitrator for quality issues for his /her organization.
Project manager	The person accountable for project success, the PM leads, monitors, and supports the project's quality management processes and procedures.
Procurement officer	Member of the project manager's team that can authorize new agreements and changes to agreements between the project team and the customer or between the project team and its vendors. The PM and quality lead work with the procurement officer to ensure quality requirements are appropriately included vendor solicitations and contracts.
Quality manager	A member of the PM's team that acts as the functional manager for quality within the PM's organization and may act as the resource manager for quality management resources supplied to the project.
Project quality lead	Member of the project organization that leads the development and execution of quality management plans, processes, and procedures at the project level.
Engineers/analyst	Members of the project manager's technical team that collect, elaborate, express, and analyze quality requirements. All approved quality requirements are then satisfied within the design of the product or service and the processes and mechanisms used to produce the product or service.
Requirements manager	Member of the project manager's team that performs the consolidation, structuring, publication, and management of requirements. May be performed by the engineers and analysts as a secondary function or by a dedicated resource. Works with the quality lead to integrate quality requirements into the project requirements data set.
Risk manager	Member of the project manager's team with training and experience in risk management. The risk manager provides guidance in the identification and analysis of project risks then manages the consolidation, structuring, publication, and management of the project risk plans and data.
Process improvement SME	A quality management practitioner with specialized training in process improvement methods such as Six Sigma or Lean Six Sigma.
Quality inspectors and testers	Members identified by the PM and the quality lead to perform quality inspections and testing as defined within the project's quality management plan.

management processes and procedures that enable the satisfaction of these requirements. During quality planning, the PM and the quality lead review all contractual documents to identify quality factors associated with each product or service. As with all project requirements, it is important to ensure quality requirements are a complete, singular, clear, necessary, and measurable.

Quality assurance provides confidence that quality requirements will be fulfilled and quality control validates requirements are fulfilled within the final products. The methods utilized for verification and validation of these requirements are defined during this planning process. The verification and validation options here are like those used during requirements management, which includes analysis, demonstration, inspection, and testing.

As with the other project management processes, effective quality management requires objective performance measurement. The metrics are used to measure product or service quality and are based on the quality requirements defined for each product and service. These measures are used during quality assurance to ensure process quality and during quality control to verify conformance. Sample measures include process cycle times, the number of defects detected, product rejections, production rates, failure rates, and customer satisfaction rates.

The project quality lead uses the organization's quality process assets to guide the planning activities. The process assets are a source of standard processes, tools, and templates used within quality assurance and quality control activities for the project. The results of this process are documented within the project's quality management plan.

As with the other project management planning processes, quality planning is based on the team's understanding of the project at that time. The reality is that things may not go as envisioned as shown in Figure 10.2. The quality plan serves as the basis for the project team to determine the impact of these realities on the project and its quality management efforts. The quality plan should be updated periodically to account for these changes if required.

The quality management plan may be included in the CDRL for the overall project or it may be included in the CDRL for a vendor hired by the project team. In both cases the plan is submitted during the early phase of the project life cycle, is continually reviewed, and any changes to the plan are managed using the project's configuration management processes as described in Chapter 9. The project quality management plan typically addresses the following topics:

- *Quality management requirements.* A description of any quality management requirements mandated by the contract.
- *Quality management system.* An overview of the quality management system to be used by the project. This section points to the organizational process assets available to the project team as described earlier in this chapter, which may be a certified QMS or the organization's quality standards and resources.
- *Quality management terms.* Definitions of all terms along with the source for these terms. This information should be available within the QMS selected for the project.
- *Customer satisfaction.* Describes how each deliverable is deemed complete and acceptable as defined by the contract and customer expectations.

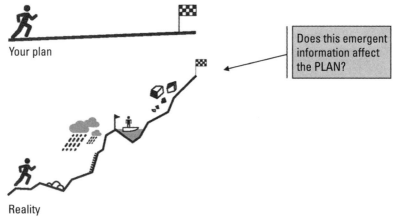

Does this emergent information affect the PLAN?

Your plan

Reality

Figure 10.2   Plan versus reality.

- *Project audits.* Defines how the project, its processes, and performers will be assessed for conformance with the organization's standards, processes, and guidelines.
- *Project reviews.* Describes the role of quality management in periodic project reviews and health checks.
- *Quality assurance activities.* A description of the quality processes, templates, checklist, tools, controls, and data to be used by the project team to prevent defects.
- *Quality control activities.* A description of the quality processes, templates, checklist, tools, controls, and data used by the project team to identify, record, and address nonconforming products and services.
- *Roles and responsibilities.* A description of the members of the project team with roles and responsibilities associated with quality.
- *Training.* A description of the quality management training to be provided to members of the project team.
- *Data collection and reporting.* A description of the quality metrics and reporting used to track and communicate the project's quality performance.
- *Inspections plans.* A description of the inspections required for incoming products and materials.
- *Document management.* An explanation of how the configuration of the project's quality management processes, templates, checklist, documents, and records are to be managed.
- *Organizational QMS.* The organization's quality policy, standards, objectives, control, assurance, and resources.
- *Quality standards.* The standards and regulations specified within the project's contractual agreement that must be demonstrated before product delivery and acceptance.

The structure and contents of this management plan can vary based on the complexity of the project and how quality is managed within the PM's organization. In all cases the quality management lead should seek to minimize the complexity of this document by avoiding duplication of information that can be found elsewhere in the project management plan or organizational process assets.

Quality Assurance

Quality assurance is the proactive set of activities performed during product development to minimize the occurrence of nonconformance. Quality assurance is first enabled by establishing a set of processes, standards, training, and mechanisms designed to enable the delivery of products and services that meet quality expectations.

Quality audits. Audits are performed on the processes and mechanisms used by the project team to determine compliance with the established standards. These process audits verify that the processes are being followed and that the processes as defined are enabling their intended purpose. During audits, the QM lead and auditors may identify potential performance shortfalls that may indicate the need for potential improvements to the processes, procedures, training, and mechanisms. These audits should be collaborative in nature, focusing on aiding the team in their consistent delivery of products and services that meet quality expectations.

Quality measurement. In parallel with these audits, quality assurance measures these processes and outcomes to determine the efficiency and effectiveness of the process. Nonconformance or customer rejection data is also collected and analyzed. This measurement data is used to indicate the potential need for improvement in the processes, mechanisms, or the training provided to the performers. Quality assurance does not specifically inspect or test the final products, but it may test incoming materials, intermediate products, work in progress, and product samples as a process assessment mechanism.

Problem solving. Issues identified by the project team through audits and/or quality measurement are resolved through collaboration between the project team and quality assurance SMEs. The team uses root cause analysis, modeling, and alternatives analysis to identify the appropriate response to problems. Any changes to the processes, training, or mechanisms are documented and used as the baseline for future audits, measurement, and problem solving.

Continuous improvement. Effective quality assurance includes a continual eye toward the improvement of process performance. There are several process improvement methodologies available to organizations. ISO 9001 calls for the continuous process improvement activities but does not prescribe the use of a specific method. The selection of the process improvement method utilized may be based on the intended outcome of the process improvement activity. For example:

- Six Sigma focuses on eliminating variability in the processes, which results in improved quality;
- Lean manufacturing's goal is to eliminate waste and reduce cost in the manufacturing process.;
- Lean Six Sigma is an amalgam of Lean principles and Six Sigma, which can be used to address waste while improving quality.

These three methods are just a few of the many available to an organization engaged in continuous process improvement (CPI). They each have unique benefits for different applications and are provided here only to introduce the PM to methods he may encounter within his organization. The key point here is that when a need to improve a process is identified, the PM should expect his quality management lead and process improvement experts to recommend and follow a known CPI framework. To help the PM understand CPI concepts, this discussion further explores Six Sigma as an example of how process improvement is performed.

Six Sigma is a quality improvement method with techniques and tools aimed at minimizing the level of variation in processes and the resulting products and services as shown in Figure 10.3. Six Sigma focuses on the voice of the customer as the basis for measuring success. Six Sigma measures process quality by striving for near perfection in the resulting product.

Figure 10.3   Six Sigma process.

The term Six Sigma comes from statistical analysis of collected data. In quality management, the data collected represents the quality scores for a set of products measured during a given period. Sigma (σ) is the Greek letter used to represent the amount of variation in the data collected. In quality management, processes are measured using the metrics defined for each quality requirement. In quality, each distribution (Sigma σ) represents a percentage of the products that met the quality expectations for the product or service. Six Sigma equates to a 99.99966 % success rate in meeting quality expectations.

The fundamental process followed when improving a process with Six Sigma includes steps to define, measure, analyze, improve, and control (DMAIC) the process.

- *Define.* Define the problem space, objectives of the improvement project, and the factors critical to quality.
- *Measure.* Understand the process and capability it provides. Measure the process's current performance. Refine the new performance objectives of the process.
- *Analyze.* Analyze the process to identify process risks and potential root causes.
- *Improve.* Determine the impact of current issues, establish acceptable variations, and implement improvements to the processes, mechanisms, or training.
- *Control.* Document the new process and enter it into configuration management.

This very high-level introduction to Six Sigma method to process improvement is an example of how process improvement experts can support the PM and quality lead.

The results of all audits and inspections performed during quality assurance are tracked within the project's quality management data set. This data set is used to measure current performance, to identify the need for performance improvement then to measure the results of a process improvement effort. Due to the specialized functionality of quality management data sets and analysis tools the PM looks to the organization's QMS for this capability.

Quality Control

Quality control involves the activities required to ensure products delivered to the customer conform to the product's requirements. The objective of this process is to identify product defects prior to delivery to the customer while achieving fast resolution of any product failures after delivery. Quality control contributes to project success by ensuring products delivered meet the customer's expectation and contractual requirements as shown in Figure 10.4. There is the risk of the customer communicating a change in expectations for product deliveries directly to the project team. These changes may constitute a change in project scope or the contractual terms and conditions. The quality lead must work with the project team to ensure

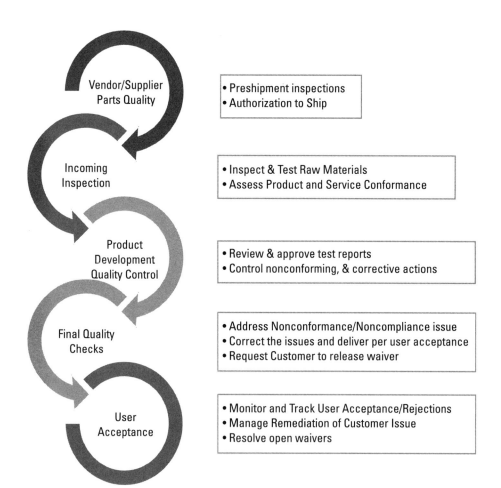

Figure 10.4 Quality control process.

these discussions are documented and vetted according the processes described in Chapters 2 and 9.

It may help to visualize the project and the project team as the factory depicted in Figure 10.5 in which products and vendor services enter through the left side of the factory and exit the right side of the factory as deliverables. Quality control supports project success by verifying the inputs and outputs conform to the project's specifications.

Vendor/supplier quality was introduced in Chapter 7 as a function used to verify and approve vendor deliverables. The quality control methods described in vendor management are defined and lead by the quality control lead. When the vendor's contract requires a predelivery inspection, the project quality lead works with the vendor's quality lead to inspect and approve deliverables prior to leaving the vendor. Any noncompliance issues are documented in a SCAR. These issues and the supplier's corrective action plan are tracked within the project's quality management data set. This data set is used for the vendor performance measurement functions described in Chapter 7.

Incoming inspections are used by the receiving organization to validate products prior to acceptance and use. The type and methods used at this stage are defined within the quality management plan. The methods utilized will vary between vendor contracts based on the type of product or service, level of preshipping inspections, and complexity of the product. As with preshipping, the quality management goal here is to ensure the vendor's products and services are compliant with the contract and specifications prior to acceptance, payment, and use by a downstream project activity.

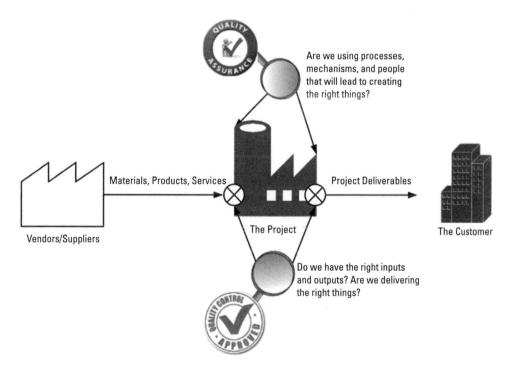

Figure 10.5 Quality control's focus.

Product development quality control involves the assessment of interim products and services created within the project team. The quality management plan may define the need to inspect products at certain increments of the production process to help reduce the impact of nonconforming items used in later production steps. The results of these inspections are documented within the project's quality management data set along with any corrective actions required for nonconformance.

Final quality checks are like those described for the vendor shipments but at this point these checks have a much higher impact on project success. These checks and inspections are the last chance for the project team to prevent nonconforming products reaching the customer. The activities associated with this step are governed by the PM's organizational QMS as well as the customer's contractual terms and conditions, all of which are documented within the QM plan. Depending on the complexity of the project and its contractual language, this process can involve a great deal of formality and paperwork. This aspect of product delivery and acceptance amplifies the importance of quality management expertise early in the project planning process to ensure the customer's quality requirements are addressed.

User acceptance includes the steps required to deliver and turn over products to the customer. As with the approval of vendor products delivered to the project, the customer may have processes and procedures for the verification and approval of delivered products. Lack of clear definition of the customer's acceptance criteria in the project contract can adversely impact the project's success and customer satisfaction. These acceptance criteria are project requirements and must be documented in the project scope, contract, and planning documents. Typically, the customer will use the corrective action requests (CARs) to document and track nonconformance issues.

Guiding principles. Six Sigma, like other process improvement frameworks, provides a set of key principles that apply here as well as during quality assurance. However, as the quality function closest to the customer, a discussion on these principles is particularly important here.

Customer focus. The term customer expectations is used throughout this chapter. Effective project quality management includes the recognition that the customer's perspective is the singularly most important perspective when it comes to defining and measuring quality.

Identify root causes. Quality management is data intensive and relies on statistical analysis to identify quality issues. Quality management also requires a clear understanding of each process and how it is intended to be utilized. However, data collection must support a clearly defined purpose and objective or else it leads to waste or inconclusive analysis. This principle demonstrates the need for a project quality management data set.

Teamwork. Quality management is an integral element of project success. The introduction to this chapter has numerous examples of how the quality lead supports other project management processes and how these processes mutually support quality management. It is important for the quality lead to be able to bring up issues in the right setting. However, adversarial relationships between quality management and the other process area performers do not lead to project success. The quality lead is a member of the project team and must act accordingly.

Flexibility and thoroughness. Quality management is a means to project success. The use of QM processes, procedures, and practitioners are the means to this end. Having quality management is not the goal. Delivering quality products and services is the goal. Regulations, processes, and procedures invoked solely for the sake of demonstrating quality management does little to lead to project success and can erode the project team's dedication to quality. Flexibility in defining the project's quality management structure will reduce the risk of having processes just for the sake of having processes.

Effective communications. Effective quality control includes clear communications with the project team about the significance of quality metrics and how these metrics relate to their work. The communication of quality metrics, such as failure rates internally and externally should be according to communication channels defined by the project manager and the PM's organization. This practice is not about hiding bad news; it is about communicating challenges effectively. Effective communication involves participation from the proper levels of the organization. Communicating issues externally should first include an understanding of the potential root causes and the potential corrective actions. It is important for the project manager and his leadership to understand all facts pertaining to nonconformance issues before communicating externally. Effective project management is contingent on effective communications.

Balanced Quality Management

Quality assurance and quality control are used in tandem as part of an effective project quality management structures as shown in Figure 10.6. Where quality assurance aims to do the right things the right way, quality control responds to correcting things that did not turn out as expected. In the daily stresses of project planning and management one may be tempted to focus energies on reacting to near term issues leaving the proactive long-term activities like quality assurance for later. The problem with this approach is that failure to allocate enough time and resources to quality assurance will most likely result in an increased rate of new issues in the form of product defects and rejections. Distractions caused by these issues further delay the ability to focus on proactive activities such as quality assurance. Relying solely on quality control will likely impact the project cost and schedule due to the time spent investigating issues, determining the root causes, and then taking corrective actions. The customer's expectation for a successful project includes the delivery of quality projects while meeting the cost and schedule constraints. Focusing efforts on quality control may lead to meeting quality expectations but it is unlikely to meet cost and schedule expectations.

New project managers should acknowledge they are now part of the organization's leadership team, which means recognizing their role in supporting their organization's success by delivering projects that lead to future business for the organization. It is important for the project manager to apply time and resources to defect prevention by understanding customer expectations and how these expectations will be achieved. By striking a balance between proactive and reactive activities the PM enables both project and organizational success by avoiding defects while meeting the customer's expectations for cost, quality, and schedule.

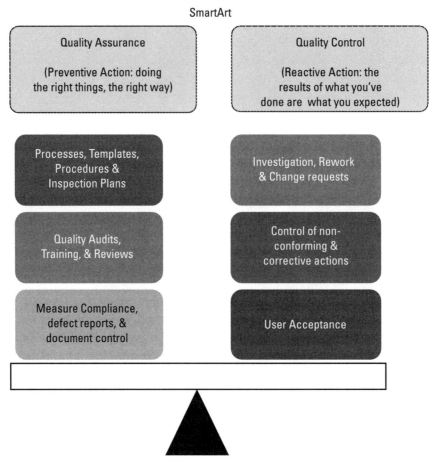

Figure 10.6   Balanced attention to quality assurance and quality control (want to shift the weight on QA).

Summary

Quality management involves the set of activities aimed at ensuring products and services delivered by the project meet the requirements and expectations defined by the customer.

Effective quality management is the result of careful integration of quality requirements, practices, and attributes into all aspects of the project and the application of quality management focused on the delivery of satisfactory products and services.

The project manager relies on members of the project team and his organization to define, implement, and monitor effective quality management practices for the project. However, as with the other project management processes the PM is accountable for the successful application of these practices. The project manager acts as the glue between the quality lead and the rest of the project team, ensuring that quality is an intrinsic element of their work.

Quality management recognizes the customer's perspective as the most important perspective for the quality of the products and services delivered. All other perspectives are secondary and communications with the customer must utilize diplomacy and tact that respects the customer's perspective.

Projects do have challenges and the potential that a requirement cannot be met in the near or long term is a reality. The events should follow the formal waiver process to document the condition and the plan to resolve the issue. The selected response to a challenge should always avoid the temptation to allow quality to be sacrificed for the sake of meeting deadlines.

References

[1]   International Organizatation for Standardization, *ISO 9001:2015 Quality Management Systems–Requirements*, Geneva: International Organizatation for Standardization, 2015.

[2]   International Organization for Standardization, *Quality Management Principles*, Geneva: International Organization for Standardization, 2015.

[3]   International Organization for Standardization, *ISO 9001:2015 Powerpoint Presentation*, Geneva: International Organization for Standardization, 2015.

[4]   Project Management Institute, Inc., *A Guide to the Project Management Body of Knowledge (PMBOK Guide)*, Sixth Edition, Newtown Square, PA: Project Management Institute, Inc., 2017.

CHAPTER 11

Tales from the Trenches

Communication Management

The Value of Informal Communications

Eric J. Roulo, aerospace structures consultant, president of Roulo Consulting Inc.

A billion dollar U.S-based manufacturer was contracted to supply a component for the Airbus A320neo aircraft. The contract involved a multilayered vendor list: Airbus contracted to a European suppler, who then contracted to Genearl Electric, which then contracted the final manufacturer.

The contract with GE limited restricted communications to the higher levels of the contracting chain. During contract negotiations with GE, it was proposed to use testing to certify their assemblies. During execution, the manufacturer would build a run of their parts, test them, and use those test results to validate the hardware for EASA (the European version of the Federal Aviation Administration (FAA)). Prior to contracting with the final manufacturer, GE had proposed certification by analysis to their prime and due to a hesitance about communicating the change of vendor and component design, GE did not communicate the manufacturer's testing approach to their customer. Instead, Airbus's European supplier submitted their analysis-based certification plan to Airbus, who promptly submitted their certification plan to EASA. By the time GE talked about the change, their European counterpart was unwilling to change the certification plan as it was already submitted to EASA. Therefore, the manufacturer was now required to certify the hardware via analysis instead of testing. The impact of this change was devastating. The cost of the project was increased by millions of dollars. Early material testing results showed design problems that were not identified during the analysis and almost held up aircraft production, flight testing, and customer delivery. All previous goodwill between the two organizations was lost at this point. The technical difficulties were eventually overcome and quality parts were delivered and installed in the airplanes. All future part testing (as originally planned) was successful with no issues. Had the vendors allowed a more open communications channel, development costs would have saved millions, aircraft assembly and flight testing would not have been delayed, and a substantial amount of good will would have been maintained between the suppliers.

The lesson learned on this project is that some informal communications across organizational lines can be useful and appropriate. It is always important to protect the private information associated with each party in a multilayered contractual arrangement. It can also be detrimental to the project if informal communications are used to direct changes to the project scope. However, informal communications within the contractual boundaries can be an effective means to solving problems.

Scope Management

Three Perspectives of Scope Management from a Single Program

Richard "DJ" Johnson, senior vice president at Booz Allen Hamilton

While supporting the U.S. DoD, I had the rare opportunity to view scope management on a single program from three different perspectives. Understanding these differing perspectives served to teach me several valuable lessons about scope management.

The program involved the development of a new system that was to operate as part of a large network of existing systems. Each of these systems were independently managed and operated but when combined provided a new capability to the DoD user community. This arrangement of systems is often referred to as a system of systems (SoS). Associated with this program there was Agency A, which was charged with facilitating interoperability of the systems comprising the SoS, and there was Agency B, which was charged with delivering the new system. The third party was Contractor C, which held the contract with Agency B to deliver the new system. There are a few important nuances associated with these parties: Agency A did not have the resources or authority to add new requirements to Agency B's project. Therefore, Agency A relied on influence and good will to achieve much of its objectives.

As the program office for the new system, Agency B was chartered to deliver a system that satisfied the requirements for a specific set of DoD users. The program office's primary responsibility was to deliver this new capability to its customers within its cost, schedule, and performance parameters. Interoperability with existing systems was important but it was not part of the key acceptance criteria in its agreement with the customer nor was it in Contractor C's contract.

Contractor C was committed to delivering a solution that met its contractual requirements but as a business, it was also had an expectation of performing its work profitably. It is important to remember that all companies are in business to make money for their owners or stockholders. Project managers employed by these companies are expected to deliver quality solutions that satisfy customer expectations while earning profitable revenues for their employer. The contract type for this project was firm-fixed price.

In the beginning of this project I was hired by Agency A to assist in influencing interoperability between the new system and the existing systems in the SoS. I started by assessing the design of the new system to determine the level of interoperability that could be expected with each existing system. My findings were not favorable. What I found was several of the existing systems would require modifications unless the new system's design was modified to support communicating with the existing systems. On behalf of Agency A, I communicated these new requirements to Agency B and Contractor C. Their response was that Contractor C did not intend to meet my request.

This was my first lesson learned: contract matters.

Contractor C's perspective of scope management was that they had a firm fixed-price agreement to deliver a clear set of requirements that did not include my requested additions. Accepting these changes would jeopardize the contractor's ability to meet the contracted requirements while maintaining profitability.

Agency B's perspective to scope management was that the requirements defined within their charter from the customer is all they were funded to deliver, and these new requirements were not part of their program funding.

Agency A's perspective to scope management was that while these requirements were important, the agency had no authority to change Agency B's scope. Their only recourse was to obtain funding to update the existing systems to support integration with the new system.

Several years later, I was hired by Agency B to assist in managing the same project. At this point, the PM for Agency B was a little concerned about the contractor's ability to meet some of the system's performance requirements. He had also come to recognize the value of implementing my previously suggested requirements but still did not have the funds to authorize a change to the contract. He directed me to document each case in which the contractor was struggling to meet the requirements and to identify the risks, puts, and takes associated with each requirement. Each time the contractor requested relief on a requirement, their waiver was denied. The PM was not trying to be difficult; he understood that his role in scope management was to ensure that each contract change was approved only after understanding the full impact of the change. By saying no initially, he was able to delay his scope decisions until he had a better understanding of the impact the changes would have on the program.

This leads me to my next lesson learned: Use no to get to yes.

The PM then began to schedule periodic project reviews focused on the complied sets of contractor waivers. The PM's strategy here was to get what he now saw as his priority requirements while allowing the contractor to successfully deliver what could be accomplished in a fair and equitable manner. The PM did this by coming to the meetings prepared to negotiate. He openly communicated his critical needs and the areas in which he felt he could relax the requirements. This openness and flexibility encouraged the contractor to be equally open and flexible in the negotiations, resulting in a win-win resolution to the scope challenges. By using patience and an initial no for each requested waiver, the PM was able to look at the project scope in a holistic perspective. Had he addressed each waiver individually, he may have failed to see the full impact of the change. More importantly, he would have given up his ability to negotiate for the new requirements.

The last lesson in scope management I wish to share: Learn how to say no.

Another area of scope management is managing customer expectations for the project. Throughout the project life cycle, there will be formal and informal engagements with the customer and its stakeholders. During these discussions, the stakeholders are encouraged to provide feedback on the system design and development. While these people do not come to the meeting with the project scope and requirements memorized, the PM must be prepared to identify potential changes to the scope. During these conversations, suggested improvements or changes to the system will be shared by the stakeholders. The PM manages expectations by politely reminding the stakeholders of what is and is not included in the project scope.

I once worked with a PM who had saying no down to an art form. He would firmly but politely say, "That is a good suggestion, but it is not in our current baseline and we will be glad to explore the feasibility of this change to the baseline." This PM knew the importance of obtaining and respecting stakeholder input, but he also knew that failing to address the suggestion as a change in scope could be interpreted as agreement to accept the change.

The Mars Viking Lander Project

Fred Brown, graduate director at Loyola Marymount University (retired)

Many of the most valuable lessons I have learned about systems engineering and project management have been learned through the challenges experienced on projects. As the author reminds us throughout this book, PM processes must be integrated to succeed, and one project comes to mind that reflects the importance of understanding and using project management processes in all aspects of one's work. While this project did occur many years ago, the lessons learned apply today as much as they did then. The Mars Viking Program consisted of two space probes, Viking 1 and Viking 2. Each probe was comprised of the Orbiter Module, which collected data and images while orbiting the planet, and the Lander Module, which collected data from the planet surface. Among the scientific objectives of this project were

- Obtain high-resolution images of the Martian surface;
- Characterize the structure and composition of the planet's atmosphere and surface;
- Search for evidence of life on the planet.

In addition to its navigation, propulsion, and communications equipment, the lander was equipped with an extensive array of sensors, collectors, and analytic capabilities to support the scientific objective of identifying evidence of life on Mars. Two of the core analytic capabilities were a gas chromatograph mass spectrometer (GCMS) and a miniature biology lab instrument. As the project scientist for the biology instrument, I was one of over 1,000 people engaged in the development of four biology experiments, each designed to look for evidence of the planet's ability to support life. These four biology experiments were chosen from the 150 proposals submitted to NASA.

During the initial planning stages for the Viking Program, the criticality of the biology instrument to the Viking Program was set so high that if the instrument was not ready for the mission, the probes would not be launched. The launch dates for the probes were set for the summer of 1975 and NASA had no intention of allowing those dates to slip. Each of the four biology experiments sought to explore unchartered aspects of the planet surface. To add to the technical challenge, the size of the module was limited: .027 cubic meter in volume and 15.5 kilograms in weight, the size of a car battery. Based on these constraints, it is no exaggeration that the probability of risk occurrence for the biology module was close to 99% and the risk consequence was a failed program.

To recap: we had been charged to build a set of unprecedented experiments to be conducted on the surface of Mars, within a container the size of a car battery, and if we failed the entire program was in jeopardy.

Lesson learned # 1: A risk not identified cannot be managed.

The technical risks for this program were great, extremely challenging but not insurmountable. Some risks were not understood or identified early in the program. If we had ignored or minimized the project's identified risks and complexity, we would not have been able to identify when things were not going as planned.

We started our work in the fall of 1970 and were slated for a preliminary design review (PDR) in the summer of 1971. As we approached the PDR, we determined that we needed more time to revise the designs to support some recent clarifications to the requirements. Therefore, the PDR was delayed until October of that same year.

Lesson learned # 2: Program reviews should only start when the deliverables are ready.

Had we held to the scheduled date of the PDR, we would have been unprepared, our design would have raised numerous unanswered questions, and the program would be delayed until the issues were resolved.

During the PDR, issues were identified with the complexity of the system design and the projected budget had increased substantially. This was not good news for anyone, but by delaying the review until we were ready, the review could focus on the critical issues with the project. After exploring multiple options to lower the cost projections and reduce risk, it was determined that reducing the scope was the only viable option. The number of experiments would be reduced from four down to three. This option was selected as the best way to reduce cost by reducing complexity. This reduction in scope also provided the benefit of increasing the weight and power budgets for the remaining three experiments.

Lesson learned # 3: Don't hesitate to request scope reductions from your customer.

By being better prepared for the PDR and openly sharing our challenges with the prime contractor and NASA, we were able to discuss alternatives and reductions in scope early in the project life cycle. Had we waited until later and hoped things would get better, the remaining three designs would have continued to struggle while we continued to spend our limited funds and time on all four experiments.

I wish I could say that the reduction of scope, while a good management decision, eliminated all our challenges for the biology instrument. But the fact is we continued to struggle and experience delays throughout the rest of the project. These issues caused our project to remain on the program manager's high visibility list for a long period of time. However, being on this list was not always a bad thing. For example: as we pushed through our design challenges, issues would arise that required a decision from our prime contractor and NASA. However, the rigid lines of communication through our company, up and through the prime and then to NASA, added weeks to each request from the design team. These delays further impacted our ability to resolve issues in a time-critical environment. To remove these delays, NASA instituted a super board consisting of representatives from NASA and our prime contractor along with our company's PM. This board was colocated with the biology instrument design team and was given the authority to make decisions on behalf of the entire project team. The turnaround time on decisions was reduced from several weeks down to a few hours.

The two Viking Mars Probes did launch as scheduled in 1975 and they entered Mars orbit the following year. Both landing crafts successfully landed on the surface and the biology instrument operated as planned and played a critical role in the data collection and discoveries made by this mission. We set out on this project to learn more about the planet Mars; little did I know how much I would learn about project management along the way.

Schedule Management

Don Macleod, chief executive officer at Applied Motion Products

Gantt charts as a staple of project planning was introduced in the United States during the First World War for the purpose of managing the fast ramp and immense scale of munitions delivery required for global military operations. Since then Gantt charts have been widely adopted and computerized, becoming the ubiquitous project planning tool used today. My experience with product planning often involves the delivery of complex technical products with a high degree of execution risk. Under these conditions management of schedule becomes a challenge. It has been said, although the attribution is not exactly clear, "Predictions are difficult, especially if they involve the future." Likewise, Gantt charts are subject to the same difficulties in predicting the completion of a technical project.

A recent case that comes to mind is the development of a step motor product for a customer in the oil and gas industry. As you might imagine, a product for use in this environment may be required to operate in the presence of hazardous or explosive gases. To ensure absolute safety in these environments, products are subject to a very intensive testing protocol to Atmosphere Explosive (ATEX) standards. Meeting these test requirements presents significant technical design challenges, especially in the case of a motor with a rotating member within the structure. A critical requirement of these products is the prevention of ignition hazard. In the event of ignition within the product, ignition gases must be prevented from escaping and posing a danger to the surrounding environment.

We embarked on this development with purchase order in hand, which means the customer was expecting delivery. We did not have the luxury of developing this product in a skunkworks environment with relatively light schedule pressure. While the development represented increased motor sales for us, each motor represented a rig installation for our customer which in turn increased daily oil production for the end user. Consequently, schedule pressure was enormous.

With due attention to detail we developed our Gantt chart and plan, laying out the key qualification milestones relevant to the ATEX standard along with the dependent tasks that would be initiated by each successful completion. With visibility to each step of the testing and planned stage gate reviews, progress and potential delays were visible to the entire team.

As often happens during a complex and highly technical development, there were some unexpected failures. In early testing ignition gas did escape the product, resulting in an explosion in the test chamber. In later testing, when the product was subject to pressure many times that measured during the explosion, the motor ripped apart. Each of these events sent the engineers back to their computers to find a fix and of course presented a threat to the schedule. Thanks to some great work by our engineers, these problems were overcome and the product achieved the much-coveted worldwide ATEX rating. The successful delivery of this project opened a whole new industry for our products.

Key Takeaways

When a serious problem arises, while it may appear that the program plan is shattered, it is important at this point to keep the team focused on the problem and not the schedule. Reanalyzing the Gantt chart and replotting critical path should be left for when the problem is solved; then you know where you are and you can move forward with the revised plan.

React immediately to schedule threats. In the case of our first problem with external ignition, I was in a parking lot in a different state after visiting a supplier. It was tempting to delay until my return to the office prior to convening a team meeting and starting work on this issue. We worked out a complete response by mobile phone in the time it took me to drive from the supplier to the airport. By the time I landed, the recovery plan was well underway.

Finally, don't underestimate the creativity, capability, and tenacity of your team to get back on track when faced with a seemingly insurmountable schedule delay. These challenges can be a critical point for the team to become more energized. When a project is faced with a setback and consequent schedule delay, the PM's leadership and hands-on involvement will help your project team pull through. Have confidence that it can be done.

Requirements Management

John (Jack) Conrad, director of program management (retired) at Raytheon (Space and Airborne Systems)

The examples that follow address lessons learned in defining, allocating, and verifying system requirements. This advice is apart from the necessity of having a rigorous change control process to maintain and track the requirements baseline. Many of these examples resulted in significant cost and schedule impacts. Hopefully, these narratives will not only alert you to specific pitfalls that lie in your future, but may they also convey an engineering mind-set that will allow you to avoid such issues.

May every project in your future meet all customer expectations and come in ahead of schedule and under cost.

AMRAAM: Write the Specification to Win the Contract!

The Advanced Medium Range Air-To-Air Missile (AMRAAM) was developed in the early 1980s in a fierce competition between Hughes Aircraft Company and Raytheon. The program promised to be a $7.5 billion effort ($21 billion today) to deliver over 20,000 missiles to replace the aging Sparrow. To accomplish the mission, the customer defined a launch envelop in which the missile would be able to acquire and track the target to intercept. The product's capabilities were written in the system-level specification.

The Air Force customer asked each contractor during the demonstration phase to build and launch six missiles, develop lower-level specifications, and submit a full proposal. To support the system-level requirement, the active radar in the missile would be required to acquire and track the target at a certain range. Hughes was excited about the capabilities of each of the elements of its active radar, which was a scaled-down version of the radar designed for the F-15 fighter. A decision was made to include specifications for each subelement of the radar to showcase the superior quality of the product. As a result, specifications were included for antenna gain, transmitter power, and receiver noise figure—all elements that factor in to being able to acquire a target at a range sufficient to track and destroy the target.

The result was a badly overspecified system. As the system was developed, some of the elements were found to exceed the allocated performance numbers, and others fell short. But because there were hard numbers for each element, there was little flexibility to perform trade-offs. As a result of these shortfalls, the cash register started to ring, and ring, and ring. The customer was reluctant to allow the redundant specifications to be changed as they rather liked the resulting performance promised in the specifications.

Lesson learned: Ensure the requirements baseline is complete but be equally sure the system is not overspecified. Trade-offs will need to be made during system development to achieve overall system performance while accounting for the capabilities of each element of the system. Eventually, the redundant specifications were removed, and a top-level requirement was maintained that fully supported the system requirement.

Communicate Clearly with the Contractor/Subcontractor

In the late 1990s, Hughes Aircraft teamed with Ryan Aeronautical (the company that had built the *Spirit of St. Louis* for Charles Lindberg) to develop the Global Hawk High-Altitude Long-Endurance Reconnaissance System. Hughes would develop the Integrated Sensor Suite (ISS) consisting of a ground-mapping radar and a 12-inch telescope with two cameras—an optical camera and an infrared camera. In 1997, Hughes was bought by Raytheon Company. The top-level requirement for the Global Hawk System was stated in one sentence: Deliver a reconnaissance platform with military utility at a given fly-away price. To meet the cost objective, contractors made as much use as they could of COTS products. One such item was the radar receiver.

By mid-2004, the Air Force customer hosted a meeting inviting Raytheon's competition to describe their offerings to replace the existing Integrated Sensor Suite. Although several issues led to this decision, at the top of the list was Raytheon's inability to deliver ISSs. A contributing factor in this shortfall was that elements of the ISS radar were being delivered for integration that failed to integrate. The director of the Global Hawk System Program (SPO) stated his frustration, "I do not like surprises....Surprises only raise credibility issues with the customer."

So, the moment came when all the stakeholders were assembled in one room. Of interest was a specification for the radar receiver that always passed final testing at the vendor on Long Island but failed when it arrived in Los Angeles for integration. The chief engineer for Global Hawk opened the meeting by reading the specification in question. He then asked the assembled group, "Does anyone understand this requirement?" An engineer from Raytheon piped up and was asked to describe the requirement to the group. After about 30 seconds, he stopped. "I guess I don't know what it means!"

As the meeting unfolded, it became apparent that the person who thought he understood the requirement the best was an engineer for the subcontractor. Sadly, his understanding was not in line with what Raytheon required.

Lesson learned: When a supplier delivers an item, the contractor and subcontractor must be confident the item will successfully integrate into the next higher-level assembly and perform as expected. To accomplish this objective, performance requirements must be clear, test requirements must be properly tiered, and the test method must be aligned with end use. In this case, a most basic discipline was skipped. At the beginning, the subcontractor must read back the requirements and state as clearly as they can, "This is what we understand this to mean."

Raytheon's flowchart describing the subcontractor relationship included "define performance requirements" but did not include a face-to-face review with the supplier, systems engineering, supply chain management, and quality management to ensure the requirements were fully understood.

View Requirements through the Eyes of the Customer

AMRAAM had been selected as the air-to-air missile of choice for the F-22 fighter. On one occasion, Hughes Aircraft engineers were asked to brief Brigadier General James Fain, later to be Lieutenant General Fain, on the capabilities of the missile. The briefing had been practiced multiple times, but perhaps not thoroughly enough. During the briefing, the lead systems engineer was waxing eloquently and lapsed into using a bit too much technical radar jargon. Finally, in frustration General Fain brought the meeting to a halt, when he literally pounded on the table and shouted, "Is this missile going to kill dirty, rotten, commie, pinko bastards or not?!!!" I was that briefer. It was a moment I will never forget.

Lesson Learned: Always step back and make sure you understand the top-level system requirements in the customer's frame of reference. The customer community will allow you to speak in your own terms only after they are assured that you understand the world in their terms.

Risk Management

Colonel (retired) Janet Grondin, United States of Air Force

I arrived in Los Angeles in June 2010, after a long drive across country with my family, to take over the space launch range division at Los Angeles Air Force Base in California. The team I inherited was incredibly talented, composed of numerous veterans of the launch business with an inherent understanding of the risks associated with launching rockets into space. My team included national experts in launch operations, flight safety, radar, optics, and many other areas of expertise. These SMEs brought with them experiences from hundreds of successful and unsuccessful launches that shaped their view of risk and what actions need to be taken for a successful launch to occur. This team did not want to lose any launches and each SME was covering all the bases to ensure National Security Space success.

And here I come, a newly minted colonel full of ideas from a year at Air War College where senior leaders challenged us to figure out how to meet the budget challenges of sequestration. I knew my mission was to drive down launch support costs while maintaining mission success. My enthusiastic approach was clearly in line with Air Force thinking, but to some on the veteran team, I looked more like a new colonel ready to make a name for herself by saving a few bucks and increasing risk of mission failure in the process.

This was my third assignment supporting satellite rocket launches and I understood pulling back our budget would make the team uncomfortable. In order to get everyone on the same page, I instituted a risk management board to proactively identify risks. Given the press on looming budget cuts, I expected that the risk discussions would be straightforward, so I prepared to roll up my sleeves with the engineering team and get down to business. However, what I found was a deep reluctance of subject matter experts to put their concerns on the table. I realized my team was still holding on to hope that the budget situation would turn around and we would not have to make any real changes to the system in order to keep our programs going. Within a couple of months of starting my new job, I knew I had a challenge on my hands to extract the data I needed out of the experts. Without their veteran knowledge of launch risks and the old systems we were maintaining and upgrading, we could easily understate critical risks or implement ineffective mitigation strategies, resulting in launch slips of $1M per day.

Over time, I was provided with expert risk management support and hired the right facilitator (the author of this book!), who helped me communicate the value of fully characterizing the risk of sequestration cuts to launch while devising appropriate mitigation strategies. We organized and prioritized our risks so we could hold technical discussions and drive to the root cause. Three aspects of successful risk management in this environment were key. First, my chief engineer understood the importance of risk and fully participated in the process, helping to create a safe environment for experts to speak up. Second, once the data was on the table, my risk manager helped the team fully characterize risks while devising effective mitigation strategies. Third, key stakeholders from the operational community participated in the process, helping to define if/then statements and validate mitigation plans. It all took a lot of patience as we wrote and rewrote risks every month until we got the full picture captured and put resourced mitigation plans in place.

I learned through this process that each one of my veteran engineers were painfully aware of many of the risks we identified based on their own experiences and design expertise. But the budget challenges cut across many engineering disciplines, and contrary to past practice, we did not have enough funds to cover every risk in every engineering discipline. Instead, we had to consider all the risks in order to apply our precious resources to the most effective mitigation strategy. We experienced some short-term failures but ultimately set up long-term efficiencies, enabled new technology to be introduced onto the range, and reduced the cost of the range operations while increasing launch capacity.

My tales from the trenches takeaway: Addressing risk is uncomfortable for many people but successful program outcomes depend on the PM's focus on defining and mitigating program risks.

Resource Management

Tejas Gandhi, chief operating officer at Baystate Medical Center

Project: Transition Hospital Operations to a New State-of-the-Art Facility

Early on during my career as the assistant vice president: Lean Six Sigma & management engineering for a major U.S. health-care provider, I faced one of the most challenging projects in my career. Our health system was building a new, state-of-the-art hospital to replace an aging and no-longer-effective facility. I was charged with leading the transition of all health-care services between the two facilities.

In health care, a retiring health care facility must be operational and ready to provide care until the last patient has been moved and the new facility must be equally operational as soon as the first patient arrives. As patient lives were at stake, our project had an extremely low risk tolerance. This readiness challenge meant the resources levels had to match the patient levels at both locations until the transition was complete and the retiring hospital was no longer in service.

Each segment of health care is very specialized, as the patients, procedures, and equipment used in one area do not always mimic those used in another area of health care. For example, the knowledge, skills, and abilities required for neonatal care do not specifically match those of the emergency room.

Health-care professionals invest years into the specialized training for their chosen field of practice and then continually learn and adjust as operations require. Therefore, it was infeasible to augment the existing resource pool with temporary resources from other health-care networks and hospital staffing levels could not support running concurrent shifts for extended periods.

Therefore, the staffing levels had to gradually ramp down as the patients moved out of the retiring hospital while ramping up at the new location to match its patient levels.

An additional challenge was that this new hospital was adopting an entirely new approach to patient care, which meant all processes, procedures, and systems used for patient care were changing. The human resources were not just moving, they were changing their entire system of care delivery.

People had to be trained and proficient on the new processes and systems while continuing to fill staff rotations at the current hospital. They effectively had to learn and be ready on the first day of a new job while still working at the old job. In patient care, there is very little tolerance for on-the-job training.

The challenges did not end at the human resources. Being concurrently operational in two places called for redundant sets of supplies, systems, and equipment.

The actual relocation steps required planning for all potential scenarios such as road closures, mass casualty incidents, systemic failures at the new hospital, and service disruptions at the retiring facility.

The project required a detailed plan that addressed

- Preparing the new hospital's facilities, systems, and equipment to be operational before transition;
- Maintaining the same high level of patient care at the retiring facility until it was no longer in service;
- Minimizing the period in which both hospitals were concurrently operating;
- Preparing employees to operate the new health-care system (people, processes, facility, and systems) well in advance of the move date;
- Providing an orderly transition of patient care between facilities that minimized all risks to patient care;
- Supporting the cost and schedule constraints defined by the hospital system.

Understanding these goals was instrumental in developing our project management strategy, which called for a set of projects, each reflecting phases of the program:

Designing: All projects that supported the design of the facility, the systems and the new processes;

Building: All projects that involved acquiring and establishing the facility, the systems, and the processes;

Commissioning: All projects that involved preparing, testing, training, and certifying the facilities, systems, processes, and staff;

Migrating: The projects to plan and execute the relocation of operations to the new hospital and then decommission the old hospital.

Lesson Learned #1: Breaking this Overwhelming Project into Smaller Pieces Reduced the Complexity

Breaking this very complex project down into smaller projects helped us to prioritize the work and better understand the resources required for each project. We still had a very complex project to execute, but we now had the ability to focus on the work rather than the challenge.

Lesson Learned #2: Clear Requirements Enabled the Resource Providers

Segmenting the work improved our understanding of the resource loading, enabling us to better communicate our resource needs to the departments. In many cases, similar talents were required on multiple projects. Working with the departments, we then identified which subprojects could not be resourced concurrently with hospital staff. However, this improved scope clarity helped the departments to identify the tasks that did not involve direct patient care and could be safely outsourced. Using external resources in lower risk areas freed up staff to focus on the critical tasks in which they were required.

The largest challenge we faced was the need to staff two hospitals concurrently. Working with the departments, we determined the best approach was to minimize the total time of the move. What we initially saw as a transition occurring over several days, was compressed into 4 hours, which gave the departments the ability to control their resource allocations.

Lesson Learned #3: Clear Communications was the Key to Success

As with any large project, communications while planning, simulating, and executing is critical for success. Regular communications between key resources, the project manager, leadership, and our other internal and external stakeholders maintained their awareness of the project during the protracted planning and preparation stages. It is important to remember that stakeholders each have their own projects and duties that reduce their level of attention to your project. This reality means that when they are not directly engaged with the project, they are not likely thinking about your project. Providing stakeholders regular updates and opportunities for feedback serves to maintain stakeholder awareness to your project and their role in its success.

The planning and preparation for this move started many years in advance and by breaking the project up into smaller projects we were able to focus our attention on smaller tasks and reduce the potential for last-minute surprises. In preparing for the move we continually reevaluated our plans and adjusted as we learned. We then tested our plans by performing dry runs of the new operations as well as the actual move. These events helped us further refine our plans while maintaining stakeholder engagement. By planning, evaluating, and replanning, the final move went off without a hitch.

Vendor Management

Rich Fitzgerald, chief operating officer at SMTC Corporation

In my experience, there is a tremendous amount of waste in the vendor management process starting with the RFI/RFQ and continuing through execution and delivery of critical to quality products. Poor vendor management results in lost time, wasted material, deliverable rework, and poor quality, all of which results in wasted money, which ultimately can lead to project failure. It is my experience that procurement officials place a tremendous amount of fluff in their schedules to ensure success in the contract award timeline and in doing so create unduly long lead times in the vendor management process. Additionally, those involved in development of procurement specifications fail to perform adequate upfront work when preparing the solicitation and then fail to communicate the project clearly to the vendors. Below are key factors that affect vendors when supporting primes:

- Lack of data in the product specification relating to the project back ground, intended use of the products and how this procurement fits into the big picture;
- Lack of clarity of scope;
- Lack of upfront scrubbing of the bill of material (BOM);
- Failure to understand the value of managing long lead times in the supply chain;
- Lack of a process or strategy for addressing end-of-life parts;
- Failure to recognize the cause and effect of product counterfeiting and the importance of preventing product counterfeiting;
- Lack of demonstrated manufacturability of the products to be supplied;
- Failure to address required testing capabilities within the scope;
- Lack of critical to quality requirements and source inspection requirements;
- Lack of overall support to quick resolution to manage timelines;
- Internal cross-functional politics of ownership to help the vendor can gain clarity.

The above are quick but important assessments I found cause delays in programs because the upfront work was not conducted.

I remember being hired by one of my prime clients to support a large-scale, 3-year, $20M program for the U.S. Navy and U.S. Air Force. By the time the program was completed, it was a $30M program and took 4.5 years to deliver. The changes in scope were so massive and so expensive, it became a grind to get the expectations narrowed down to a set of objective acceptance criteria. The prime contractor allowed the end client to make many changes and did not lock down the design. The prime also failed to follow a structured change management process. The prime further exacerbated these changes by failing to flow these new expectations to its subcontractors. However, the prime still expected the vendors to meet these undocumented changes in expectations. Changes and renegotiation of SOWs and contracts after the change has been accepted by the prime are very complicated, require executive attention, and put avoidable stresses on the buyer/supplier relationship. I found that once we stopped, assessed, and presented the cause and effect, our prime client understood the impacts of these changes on our scope.

The following are a few lessons I have learned in my career that can help to avoid waste on your projects:

Vendors often try to please the client by focusing on speed, and unfortunately speed in change management can kill a project. You must stop and assess change. Although I understand the pressures of time, failure to document any changes in scope to save time will ultimately impact the overall success of a program.

Inventory and supply chain management are equally important steps in the vendor management process, both of which have risks that must be adverted. Often due to copy exact modeling, end-of-life becomes prevalent in the quality designs.

Always buy from registered distribution or follow the Government–Industry Data Exchange Program (GIDEP) processes to avoid counterfeiting. Companies who have strict counterfeit programs typically will work better with primes.

Test requirements are key to supplier partners who want to ensure that the product will outlast the life of the design rules. Primes who work closely with supplier partners to perform rigorous subassembly testing at the supplier site typically show lower field failure results.

If a customer wants you to win, you win. Recognize upfront that suppliers like a prime are not perfect. Work with your supplier for equal success, and you will find a much more motivated workforce, ultimately making you more successful.

What gets measured gets done. Ensure your supplier partner is aligned to the metrics you are being measured against. Then ensure they measure themselves accordingly inside their factory at the tactical level, which rolls all the way back up to your success metrics.

Reward successful behaviors. Build into your budget a rewards system for the companies and factory employees you are working with! They will forever be allegiant to your success. A small act goes a long way. When commending your suppliers, bring out the program manager and the end customer; this small act will get the supplier emotionally attached to your product and success.

Cost Management

David Kubera, senior program manager at Defense Aerospace Corporation

The EVMS is one of the most essential tools available to the PM for controlling costs no matter the size of a program or project. The EVMS is a way to tell if you are winning or losing on a program from both a cost and schedule point of view. If you are losing (overbudget or behind schedule), you can perform a variance analysis to help you understand the root cause and how you can get the project back on track. The EVMS can be fully compliant to the ANSI Guidelines 32 Criteria or as an EV-lite level depending on your program requirements. I've experienced various scales of EVM maturity and have seen what works and what doesn't. The following are keys to implementing a successful EVMS.

Planning

It is imperative to apply a deliverable-oriented WBS structures from the start of project planning. Too often, individuals plan from their functional perspective, not from the perspective that a deliverable requires multiple functional areas that work interdependently together to complete the design and development of subproducts that make up the final product. It is difficult to measure performance of a function rather than measuring the status of the design of each subcomponents that make up a system deliverable. Think of a bicycle that is made up of wheels, a frame, handle bars, and so on. All these subcomponents require detailed tasks to be completed before a bicycle can be assembled, tested, and delivered. Cost and schedule plans organized by functional resource area do not help the team know where they are in a project.

Planning at the Right Level

Planning your work packages to a level that is manageable is critical to an effective EVMS. I have seen where planning is at too high of a level to provide meaningful progress indicators to the team. I have also seen where each operation on a shop floor is detailed in a schedule and value is earned at that level. Although intentions are good and the schedule is detailed and thorough, it is not a practical approach. The level of tracking planned must also be supportable by the scheduling and financial analysis resources supplied to the team. There needs to be a balance of watching the scoreboard and playing the game.

Culture of Accountability

In order to implement an effective EVMS, it is imperative that all stakeholders understand what EVMS is and what processes, procedures, roles, and responsibilities are required. Resources should be dedicated to documenting an EVMS description and processes so that there is clarity on what everyone needs to do. Documented within the responsibility matrix, control account managers (CAM) are assigned to subelements of a WBS structure and are accountable to the cost, schedule, and technical performance of that subelement. CAMs typically have a day job (i.e., functional engineer) and the CAM role is perceived to be a secondary responsibility. There is never enough training that can minimize the perception of EVMS as a burden. Nevertheless, be armed with the benefits of performing EV on a project or program and prepared for objections.

Monitoring Cost and Schedule Management

Louis Gerstner, former IBM CEO, said, "People do what you inspect, not what you expect." Having an EVMS system is one thing but having periodic reviews with your team where they report to you on cost and schedule progress is essential. In those reviews, you can monitor your team's attention to schedule and budget. You have an opportunity to discuss strategies on how to get back on track. Perhaps you are realizing savings under one control account that could offset overruns on another control account. You can relook at your requirements and verify that you are not expending extra cost and schedule to overachieve on a requirement. Perhaps you need support to gain resources that are more efficient than your current resources. Lastly, never underestimate the value of managing the risks and opportunities. Putting focus on executing risk countermeasures within your baseline will dramatically reduce the risk of budgets overruns. Leverage the opportunity register to harvest savings from activities such as design to unit production costs (DTUPC), bundling material, or performing continuous improvement events.

The Importance of Cost Awareness

Larry Johnson, president of LR Johnson Associates LLC

As an experienced management consultant and business owner, I have been involved in a wide variety of projects, both as a manager and as a consultant to management.

These projects have spanned across a broad spectrum of industries. As I look retrospectively at my experiences and explore lessons learned, a few points arise as noteworthy:

Establish the project vision and obtain team buy-in. It is critical to develop a clear vision and definition for your project, and then get buy-in and commitment from your team. The project vision should always include cost, schedule, and performance attributes. Have your team members share—openly and without threat of retaliation—their concerns and apprehensions about the project and how they might address their concerns. My experience suggests that without a shared vision and consensus on the goal and process to move forward, your project may suffer. Without this first basic step it is easy to lose focus and become distracted as things change in the environment. If there is anything that experience tells us, it is that things will happen, and it is important to consider the wisdom and lessons learned by your team as you develop and implement your project plan. While this approach to stakeholder input can be fraught with challenges, a good manager should be capable or managing varied input and evaluating the merit of each issue, incorporating all feedback in the project plan.

Align to your organization's values. Assess your organization's value drivers. (i.e., those qualities that your team possesses that will enable you to accomplish your goals and achieve your project's vision). Also identify the qualities of the project itself that align to the organization's values. This exercise will enable managers to identify any needed skills or insights necessary to satisfy the vision and to appreciate the role cost plays in project success. The managing partner in the consulting firm I once worked in had a mantra: "Surprises are for birthdays." Eliminate the risk of surprises in your project by first insuring that you have the tools and insights necessary to successfully complete your project in place.

Balance cost, schedule, and performance. Successful projects deliver products and services that satisfy the customer's requirements and are delivered within the allotted time and budget. A project that exceeds customer expectations but is 2 years late and cost twice as much as budgeted is not likely to achieve accolades.

Effective project planning requires balanced attention to cost, schedule, and product performance. Effective cost management requires clearly delineated financial objectives. Cost management gives us a framework to monitor performance and gauge progress while reducing the potential for surprises. Such a detailed plan requires cost benchmarks against which progress can be measured. This is where my earlier suggestion comes to play; the project plan and its benchmarks should be reviewed and agreed to by the team. Financial data is frequently reviewed against these benchmarks. These periodic touchpoints allow for a review and analysis of cost and schedule results. Such a review should not just determine whether objectives have been met but understanding the implications of the results and the numbers behind them. Are these results sustainable? Are the data points being propped up by unnatural activities by managers focused on achieving near-term goals at the expense of long-term organizational health? What are the trends? Can these results be expected to continue? I believe it was Samuel Clemens (Mark Twain) who said, "Figures don't lie, but liars figure." Review results with some skepticism. There have been recorded instances of managers who, due to organizational pressures, manipulate results to show desired outcomes. Analyze cost anomalies, including perfect results, to make sure they explainable. Special attention should also be paid to cost overruns, as they may symptomatic of underlying problems.

Effective cost management is more than monitoring results against the budget. Effective cost management requires recognition of the importance of cost by the entire team.

Quality Management

John D. Lillard, founder of Capability Assurance Institute

As a project manager over the past 25 years, I did recognize the role of quality in project success and how consistently meeting customer expectations led to future business. However, I admit that I have not always recognized the importance of ensuring quality was baked into the system specifications and design. During my 17 years of supporting the U.S. Navy as a civil servant, I was frequently brought in to help troubleshoot struggling projects. The challenges these projects faced were rarely due to the skills and expertise of the project team, as some of the best and brightest people in the country were working on these projects. What I brought to the table was a fresh set of eyes and the ability to achieve common ground between the customer and the project team. The example I wish to share fits the case of a project full of hard-working people who cared deeply about their work, but for some reason could not get the project over the finish line.

In 2007, I was asked to help the U.S. Navy Anti-Terrorism and Force Protection (ATFP) program with a project that was technically complete, but the end user would not accept the system due to several concerns he had with the quality of the solution.

The objective of this project was to elevate the level of service for 911 and emergency dispatch provided to Navy bases in the aftermath of the attacks of 9/11 and the shootings at Fort Hood. This was a very complex project that involved building new dispatch centers and equipping them with state-of-the-art systems for 911, alarm monitoring, radio communications, and computer-aided dispatch. The project also involved the installation of radios and computer terminals in every Navy fire truck, ambulance, and patrol vehicle.

The project plan called for the maximum use of commercially available products with proven use in use in the public sector. This strategy was meant to reduce the technical risk associated with the project while reducing the life cycle cost of the installed systems. Due to the high level of commercially available products, the need to create detailed engineering documentation was relaxed. All these tactics to improve the quality of the solution seemed reasonable.

One of the customer's concerns was with the quality of the in-vehicle computer installations. The issue was that the computer terminals once installed in the vehicles often wobbled on their mounts and the cable connectors were prone to come unplugged. The root cause of the wobbling was traced to the method used to connect the mounting brackets to the vehicle floor. The floor, not the mount, was flexing under the load. It appeared that the contractor failed to account for the total weight of the bracket and computer in relation to the floor's ability to hold the weight. When I asked the subcontractor how this happened, he stated that the government could not provide specific data on the vehicle construction and the government engineer approved their designs. He also stated that this could have been avoided by using lighter computers. When I asked as to why such a heavy computer was selected, the prime contractor stated that it was the only one that met the Mil Standard specified by the government. When I inquired with the government engineer as to the reason for the spec, he stated it was the only one he could find that defined the level of ruggedness requested by the end users and the alternative would have been to write his own specification.

When I investigated the issue with the connectors, I learned that the PC selected did not support ruggedized connectors, which could have prevented this issue. I already knew why the PC was selected. This story demonstrates how a series of minor decisions can lead up to a major problem:

- The Mil Standard was picked to ensure that the laptops would meet the physical demands associated with vehicle installations. This seemed to be a sound way to ensure quality. However, the Mil Standard specified a level of ruggedness beyond what was required for this project. The specification limited the solution space and resulted in a difficult product to install. Lesson learned: Overspecifying quality in one area may negatively impact quality elsewhere.

- The vehicles receiving these computers were not yet delivered and the third-party provider (GSA) stated that the exact make and models could not be provided until the time of delivery. As the brackets were custom products, the schedule did not allow for waiting until the vehicles arrived. The concerns raised by the systems engineer about this lack of information went unaddressed by the program manager. Lesson learned: Quality solutions are those that accurately address the intended environment. Not having all the information about the environment is likely to impact the quality of the solution.

- The installation contractor provided detailed designs of his mounting brackets based on the estimated loads from the computers. But he could not demonstrate the performance of his brackets once installed within the vehicle as there were too many vehicle variants. Lesson learned: Subsystem designs require demonstration of the subsystem installed and in use.

- The government engineer did validate the subcontractor's weight calculations for construction of the mount. But the designs could only show a generic vehicle for mounting, which could not be verified for accuracy. Lesson learned: Always ask, What information is missing from the design, plan, or document, and should I be concerned? In this case, the solution was a heavy plate under the floor to add stability. This change came after a significant delay, additional cost, and damaged customer confidence.

- The computers and the devices connecting to them were both commercially available products with industry-standard connectors. Therefore, there was no requirement to provide detailed diagrams for these connections. Without drawings of these connections, the durability of these connections was not questioned. The issue here is that the project team did perform extensive testing of the terminals in a lab environment. The tested configuration reflected the planned installed configuration. However, the testing failed to validate that the quality requirements were being met. In other words, the issue was right in front of all of us, but we were not looking for quality issues. Lesson learned: Quality solutions require an eye for quality at every step of the project.

In hindsight, these issues may seem obvious and avoidable. But prior to this project, many of us would have done things the same way. I also see now the value of having a person who is focused on quality as a full-time job. A good quality lead asks the questions unthought of and sees things in a perspective different from us. I hope your future career in project leadership brings you the same level of challenge, satisfaction, and education as it has for me.

Configuration Management

John (Jack) Conrad, director of program management (retired) at Raytheon (Space and Airborne Systems)

The AWG-9 Radar for the F-14 Aircraft: A Persistent Test Failure

When I think back on my 40 years of developing high-tech weapon systems, there is one problem that always seems to be at the top of the list of the most difficult problems to solve. The radar for the F-14 fighter consisted of several individual units such as the antenna, transmitter, and receiver. The unit that seemed most difficult was one that everyone thought would be one of the easiest ones to build. The unit was called the Synchronizer and incorporated a crystal oscillator and associated circuitry to generate the timing signals for the radar and the microwave frequencies utilized by the radar.

The radar units were built by Hughes Aircraft Company in Los Angeles and delivered to the prime contractor, Grumman Aircraft, on Long Island. During manufacture, the unit was calibrating and tested by performing acceptance testing on an (Grumman) AWM-23 test station. These Grumman unit test stations would eventually be delivered and installed on various aircraft carriers to allow the radar to be maintained by the Navy.

Grumman had developed their own test stations to test the units prior to installation on the F-14 aircraft. An issue arose when units that passed acceptance testing in Los Angeles consistently failed when tested in Long Island. The unit would be returned, only to pass on the west coast. When the unit was retested on the east coast, it failed. After a while, people were joking that the units liked California better than New York, but it was no laughing matter. The problem was bringing the program to its knees. The most experienced engineers available were assigned to get to the root of the problem. Month after month passed, and unit after unit failed. There seemed to be no end in sight.

The problem continued for nearly a year when finally, an engineer from Hughes Aircraft spotted the problem. Unbelievably, after one person spotted the cause of the failure, no one could believe it had taken so long. For the synchronizer unit, the AWM-23 test station used in Los Angeles had a pedestal about 18 inches tall with a connector at the top of the pedestal. The unit was dropped down vertically to allow the unit connector to mate with the test station connector. The Grumman Long Island facility mounted the unit horizontally (which, by the way, was also the way the unit was mounted in the aircraft. The failure was caused because the two test facilities tested the unit in two different orientations. At the heart of the problem, the crystal oscillator that was used as a frequency reference was enclosed in a casing that included a small heating element to ensure the crystal was kept at a constant temperature, since the frequency of the crystal changes slightly with temperature. The transfer of heat from the heating element to the crystal turned out to be dependent on unit orientation. When the units were tested in both facilities in a horizontal orientation, the units consistently passed on both coasts.

Lesson learned: The requirement to test units in a manner that reflects end use is seldom given a moment's thought. After all, test is testing. In this case, it made a significant difference and failure to follow the end use conditions led to sizeable cost and schedule impacts. By the way, the engineer who spotted the issue received a nice addition to his paycheck!

About the Author

Nehal Patel (PMP) is an author and program manager with 17 years of aerospace experience using engineering-based critical thinking to execute programs for a variety of complex systems involving corporate, government, military, and vendor relationships. She has experience with each stage of the product development cycle from conception and corporate through long-term sustainment and production management. She is currently a senior program manager at a multibillion-dollar aerospace and defense company. She received a B.S. in electrical engineering from University of Massachusetts, Amherst, and an M.S. in systems engineering and Leadership Programs from Loyola Marymount University, Los Angeles. She is also PMP-certified from the Program Management Institute.

Index

A

Abbreviations, email, 9
Acceptance criteria, 20
Accessibility, 12
Accountability, culture of, 212
Activities
    constraints of, 48
    defining, 37–43
    dependencies of, 43–44, 48
    requirements elaboration, 72
    resource estimation, 52–53
    sequencing, 43–45
    See also Scope management
Activity duration
    analogous estimating, 50
    elapsed time and, 47
    estimating, 45–52
    estimation techniques, 49–52
    expert judgment in, 49–50
    PERT and, 50–51
    reserve analysis and, 52
    resource loading and, 49
    rollup, 48–49
    wait time and, 48
Actual cost of work performed (ACWP),
        162–63, 165, 167
AMRAAM example, 204, 205
Analogous estimating, 50
ANSI/EIA-748-C 32 guidelines, 151–52
Audience, knowing, 2–3
Average equation, 51
AWG-9 radar example, 216–17

B

Back channels, 11
Balanced quality management, 193–94

Baselines

Baselines
    configuration items, xix
    cost management, 149, 160–63
    performance measurement, 162
    risk management, 85
    schedule, xvii, 53–56
    scope management, 17, 22, 24, 30
BCC (blind carbon copy), 7
BLUF (bottom line up front), 8
Brainstorming, 92
Budget cost of work scheduled (BCWS)
    amount of, 166
    defined, 164
    example calculations, 164
Budgets
    with BCW, BAC, and PBB, 160
    establishing, 158–60
    network diagram and Gantt chart, 159
    project base (PBB), 158, 160
    summed to control accounts, 161

C

Calendar constraints, 54
CC (carbon copy), 7
Change management, 30
Changes
    communicating, 119
    configuration, evaluating, 179
    configuration, managing, 175–78
    example classifications, 178
    impact on cost, schedule, and performance,
        65
    management process, 176
    resource, 124
Collection
    documents, 21
    requirements, 24–26, 68–71

Communication management
    defined, xvii
    direct, informal communication, 11–12
    email use, 5–9
    examples, 197–98
    knowing audience and, 2–3
    media, 3–5, 6
    meetings, 10–11
    phones and voice mail, 9–10
    templates, 12–15
Communications
    accessibility and, 12
    back channels, 11
    collaborative face-to-face, 3
    contractor/subcontractor, 204–5
    defined, 2
    effective media, 3–5, 6
    eye contact in, 12
    informal, 11–12, 197–98
    with project team, 121
    role in project success, 1
    rules for, 1–2
Concept design review (CoDR), 41
Configuration items
    changes, managing, 175–78
    evaluating, 179
    identifying, 175
    releasing new, 179–80
    status, maintaining, 178–79
Configuration management (CM)
    AWG-9 radar example, 216–17
    change management process, 176
    configuration change management, 175–78
    configuration items identification, 175
    configuration item status, 178–79
    contractual documents and data, 173
    defined, xix, 171
    design evolution, 174
    evaluation criteria, 179
    internal data and documents, 173
    new item release, 179–80
    objective of, 171
    organization, 174–75
    planning, 172–75
    process, 172
    project review cycles, 174
    requirements data, 173

roles, 177
    summary, 180
Consequence scoring criteria, 99
Constraints
    activity, 47, 48
    budget, 73
    calendar, 54
    design, 82
    environment of, 19
    finish date, 48
    performance, 85
    resource, 92, 96
    schedule, 24, 29, 31, 50, 54, 79
    scope management and, 19
    start date, 48
Continuous process improvement (CPI), 188
Contracts
    awarding, 138–39
    cost-plus, 134–35
    fixed-price, 132–33
    time-and-materials, 133–34
    See also Vendor management
Contracts management, 146–47
Contractual conditions, 20
Control accounts, 157, 160, 161
Corrective actions, schedule, 59–61
Cost
    balancing schedule and performance and,
       213
    change impact on, 65
    effective project planning and, 213–14
    in vendor analysis, 138
Cost awareness
    importance of, 212–14
    organizational values and, 213
    project vision and team buy-in and, 213
Cost management
    culture of accountability and, 212
    defined, xix
    example, 211–12
    measurement methods and baseline defini-
       tion, 160–63
    monitoring, 212
    planning, 211–12
    in PMP, 20
    processes, 152
    project organization, 153–56

project performance analysis, 163–68
risk impact on, 86
roles, 153
schedule and budget organization, 156–60
summary, 168–69
WBS, 153–54
See also Earned value management system
    (EVMS)
Cost-plus contracts, 134–35
Cost-plus-fixed-fee (CPFF), 134
Cost-plus-incentive-fee (CPIF), 134–35
Cost-sharing, 134
Cost variance (CV), 166
Crashing, schedule, 60
Critical path, 58–59
Customer satisfaction, 186

D

Define, measure, analyze, improve, and control
    (DMAIC), 189
Delays, accepting, 60
Deliverables, 19, 29
Dependencies
    activity, 43–44, 48
    critical, identification of, 59
    predecessor, 48
    schedule, 33, 36, 39
    successor, 48
Diagramming, 44–45, 46, 93, 94
DISC assessment methodology, 2–3, 4, 5
Document review, 93
Documents
    collecting, 21
    project initiation, 67, 72
    source, 20

E

Earned value management (EVM)
    analysis values, 163
    defined, xix
    resource loading and, 49
    trends, 167, 168
    visibility and, 150
Earned value management system (EVMS)
    ANSI/EIA-748-C 32 guidelines, 151–52
    defined, 149–50

example data collection and calculations,
    165
history of, 152
principle indicators, 150
visibility, 150
visualization of data, 166
Eighty-twenty reports, 14
Email
    abbreviations in, 9
    BCC, 7
    BLUF (bottom line up front), 8
    CC, 7
    confirmation and priority features, 8
    distribution, 6–9
    distribution wars, 9
    effective subject lines, 7–8
    To: field, 6
    reading, 6
    signature blocks, 8
    as time-wasting activity, 5–6
    use of, 5–9
Expert judgment, 49–50
Eye contact, 12

F

Fast tracking, 60
Financial conditions, 20
Firm fixed price (FFP), 133
Fixed-price award fee (FPAF), 133
Fixed-price contracts, 132–33
Fixed-price economic price adjustment (FP-
    EPA), 133
Fixed-price incentive (FPI), 133
Fixed-price level of effort (FP-LOE), 133
Focal points, 40
Four corners chart, 12–13, 15
Functional organization, 111–12

G

Global WBS (GWBS), 39
Government-Industry Data Exchange Program
    (GIDEP), 211

H

High-priority risks, 99
Hospital operations transition example, 207–9

I

Informal communications, 11–12, 197–98
Initial requirement matrix, 25
International Council on Systems Engineering
    (INCOSE), 63, 68–71, 82, 172
International Organization for Standards (ISO),
    63, 73, 82, 142–43, 181–83
Interviews, 91–92
Invoicing strategy, 43

L

Lead and lag times, 45
Lean Six Sigma, 207
Level of detail, 41–42
Life cycle
    approaches, 23–24, 174
    documenting, 23
    support, 20
    system development (SDLC), 39
Low-priority risks, 100
Low-rate initial production (LRiP), 143

M

Mars Viking Lander project, 200–202
Matrix organization, 113–14
Medium-priority risks, 99–100
Meetings, 10–11
Milestones
    in PMP, 20
    reviews, 41
    schedule alignment to, 53
    weighted, 160
Mil Standard, 215
Monitoring
    cost management, 212
    performance, 30
    project team performance, 123
    risks, 104–5
    schedule management, 212
    vendor performance, 144–47

N

Network diagrams, 44–45, 46

O

Opportunity management, 105–6
Organizational process assets (OPAs), xix, 23
Organizational systems, xix–xx

P

Perception versus optics, 4–5
Performance
    analyzing, 57–58
    assessing, 57
    balancing cost and schedule and, 213
    change impact on, 65
    effective project planning and, 213–14
    measurement baseline, 162
    project, analyzing, 163–68
    project team, 123
    risk impact on, 86
    schedule and, 57–58
    vendor, 144–47
    in vendor analysis, 138
Phase gate reviews, 20, 40–41
Phones, 9
PMBOK Guide, 122–23
Point of contact (POC), 158
Procurement officer, 127, 129
Procurement strategy, 132
Product/services
    acceptance of, 141
    delivery, 140–41
    quality control methods, 142
    quality inspections, 143
    received, 141
Program evaluation review technique (PERT),
    50–51
Project audits, 187
Project base budget (PBB), 158
Project environment, 21
Projectized organization, 112–13
Project leadership, 162–63
Project management
    effective, xv–xvi
    processes, xvi–xx
    See also specific types of management
Project management plan (PMP), 18, 19

Project organization
    defining, 155
    example, 155
    integrating WBS to, 155–56
Projects
    origination of, xx
    overview, 19
    planning, xx
    sample, xx
    success or failure criteria, 2
Project Scope Statement, 27
Project success, xvi
Project team
    being vested in, 121
    buy-in, 213
    communication and, 121
    developing, 120–23
    growing talent from within and, 122
    knowing audience and, 122
    leading by example, 121
    as living organism, 124–25
    managing, 123–25
    performance, monitoring, 122–23
    teamwork, 122
Project vision, 213
Prototyping, 70
Purchase orders, awarding, 139–40

Q

Quality assurance (QA)
    activities, 187
    balanced attention to, 194
    confidence, 185
    continuous improvement, 188–90
    defined, 183
    elements of, 184
    problem solving, 188
    quality audits, 184, 187–88
    quality measurements, 188
Quality audits, 184, 187–88
Quality control (QC)
    activities, 190
    balanced attention to, 194
    communications, 193
    defined, 184

focus, 191
guiding principles, 192
inspections, 143
methods, 142
process, 190
product development, 192
Quality management (QM)
    balanced, 193–94
    defined, xix
    example, 214–15
    items, 186
    knowledge area, 183
    overview, 181–84
    planning, 184–87
    plan versus reality, 186
    processes, 182
    quality assurance and, 187–90
    quality control and, 190–93
    requirements, 186
    roles, 185
    strategy, 21
    summary, 194–95
Quality management system (QMS)
    assessment, 183
    defined, 181
    in quality management plan, 186
Quality manager (QM), 142, 143
Quality measurements, 188
Quality standards, 187

R

RACI matrix, 89
Reporting
    risk, 106
    time frame, 42
    value of, 42
    vendor, 145
Request for information (RFI), 135–36
Request for proposal (RFP), 135–36
Request for quote (RFQ), 135–36
Requirement relief, 79–82
Requirements
    allocating, 75
    analyzing, 76–82
    collecting, 24–26, 68–71

Requirements (continued)
  data set, creating, 74
  decomposing, 71–72
  defined, 63
  elaborating, 71–74
  expressing, 74–76
  extending, 72
  focus on, 120, 122
  metrics, 82
  negotiating, 79
  prioritizing, 73–74
  publishing, 75–76
  quality management (QM), 186
  quality measures, 83
  resource, 24–26
  in scope management, 24–26
  as single thread, 64
  terms, 67
  testing, 211
  tracing, 79, 80–81
  validating, 78
  verifying, 76–78, 82–83
Requirements list, 26
Requirements management
  defined, xvii–xviii
  examples, 203–5
  goal of, 63
  process, 64
  process in depth, 66
  requirements, analyzing, 76–82
  requirements, expressing, 74–76
  requirements collection, 68–71
  requirements elaboration, 71–74
  roles within, 65
  tool, 76
  user needs identification, 66–68
  verification and validation, 82–84
Requirements matrix, 75
Requirements model, 77
Reserve analysis, 52
Reserve funds, 101
Resource loading, 49
Resource management
  administration control, 110
  defined, xviii
  examples, 207–9

fragmented tasks, 116
  functional organization, 111–12
  matrix organization, 113–14
  operational control, 111
  organizational environment, 109–14
  organization constructs, 117
  organization types and, 111–14
  partnerships in, 119
  planning, 54, 114–17
  PM workload, 116
  processes, 110
  project control, 111
  projectized organization, 112–13
  project team development and, 120–23
  project team management and, 123–25
  relationships in, 119
  responsibilities of, 110–14
  roles, 110–14
Resources
  acquiring, 117–20
  changes, coordinating, 124
  dedicated, 115
  estimating, 52–53
  internal, 119
  leveling and availability, 54
  requirements, 20, 53
  support-as-a-service, 115–16
  workload, 116
  work package delegation, 115
  work rollup by type, 118
Reviews
  concept design (CoDR), 41
  milestone, 41
  phase gate, 20, 40–41
  project, 187
Rewarding successful behaviors, 211
Risk identification
  brainstorming, 92
  categories, 95–97
  diagramming, 93, 94
  document review, 93
  example structure, 97
  interviews, 91–95
  kickoff meeting, 94–95
  methods of, 91–92
  preliminary, results of, 96

process, 91
statement, 97
SWOT analysis, 93, 94
Risk management
defined, xviii
example, 206–7
guide, 85
handling risks, 102–4
information elements, 88–89
opportunity management and, 105–6
organization, 89
overview, 88–89
overview illustration, 91
planning, 88–90
as proactive, 85
process, 86, 89–90
reserve funds and, 101
risk analysis, 98–102
risk identification, 91–98
risk monitoring, 104–5
roles within, 87
summary, 106–7
tools and templates, 90
Risk register, 101
Risk(s)
accepting, 102
analyzing, 98–102
avoiding, 102
burn-down, 104
consequence scoring criteria, 99
defined, xviii, 85
exposure, 100
handling, 102–4
high-priority, 99
impact on cost, schedule, and performance, 86
initial assessment, 21
likelihood, 89, 98
low-priority, 100
management, 42
medium-priority, 99–100
mitigation, 103
monitoring, 104–5
objective of handling, 104
ranking and prioritization, 98
reporting, 106

transfer, 102–3
Roles
configuration management, 177
cost management, 153
documenting, 22
human resource management, 114
quality management (QM), 185
requirements management, 65
risk management, 87
schedule management, 37
scope management, 19
understanding, 119, 121
vendor management, 128
Rolling wage plan, 42–43

S
Schedule management
activities, defining, 37–43
activities, sequencing, 43–45
activities duration, estimating, 45–52
activity resources, estimating, 52–53
defined, xvii
examples, 202–3
monitoring, 212
objective of, 33
overview, 33–37
performing, 33–34
process, 33, 34
roles associated with, 37
schedule development, 53–56
schedule use, 56–61
summary, 61
Schedules
alignment to milestones, 53
analysis, 53
balancing cost and performance and, 213
baselining, 55–56
change impact on, 65
compression of, 60
corrective actions and, 59–61
crashing, 60
critical path analysis and, 58–59
developing, 53–56
developing with network logic, 156–58
effective project planning and, 213–14

Schedules (continued)
  evolving detail, 36
  example, 33, 35
  influences, 36
  performance, analyzing, 57–58
  performance, assessing, 57
  progress, tracking, 56–57
  risk impact on, 86
  using, 56–61
  in vendor analysis, 138
Schedule variance (SV), 166
Scope
  controlling, 29–31
  defined, xv
  documenting, 26–27
  reducing, 60
  validating, 29
Scope management
  change conditions and, 30–31
  collect project requirements and, 24–26
  defined, xvii, 17
  documenting scope and, 26–27
  goal of, 17
  Mars Viking Lander project, 200–202
  performance monitoring and, 30
  perspectives from a single program, 198–202
  planning, 21–24
  plan project management and, 18–21
  process, 18
  RACI matrix, 22–23
  roles associated with, 19
  saying "no" and, 199
  in schedule development, 54
  scope control and, 29–31
  scope validation and, 29
  summary, 31
  WBS creation and, 27–28
Sequencing activities, 43–45
Signature blocks, email, 8
Six Sigma, 188–89
Stakeholders, identifying, 21–22
Subject lines, email, 7–8
Subject matter experts (SMEs), 25–26, 49
Surge capacity, 116
SWOT analysis, 93, 94
System development life cycle (SDLC), 39

System of systems (SoS), 198

T

Task analysis, 70
Teamwork, 122
Technology readiness levels (TRL), 96
Templates
  eighty-twenty reports, 14
  evolution of, 14
  four corners chart, 12–13, 15
  importance of, 12–15
  trip report, 12, 13
  See also Communication management
Time-and-materials contracts, 133–34
Tracing requirements, 79, 80–81
Tracking progress, schedule and, 56–57
Trip reports, 12, 13

U

Use case analysis, 70
User needs identification, 66–68

V

Vendor management
  defined, xviii
  example, 209–11
  illustrated elements, 127
  performance monitoring, 144–47
  processes, 129
  procurement officer and, 127, 129
  product/service acceptance, 140–44
  relationships in, 144–45
  roles within, 128
  scorecard, 146
  summary, 147–48
  vendor selection, 130–40
Vendors
  analysis, 138
  contracts, awarding, 138–49
  defined, 127
  evaluation criteria, 137
  negotiating with, 137–38
  payment to, 144
  performance process, 130
  product/service acceptance, 140–44

purchase orders, awarding, 139–40
reporting, 145
responses, evaluation of, 136–37
shipment and acceptance process, 140
solicitation of, 135–36
Vendor selection
cost-plus contracts, 134–35
fixed-price contracts, 132–33
offer evaluation in, 129
process, 130
procurement strategy, 132
steps, 131
supplier requirements definition, 131
time-and-materials contracts, 133–34
Voice mail, 9–10

W

WebEx, 10–11
Weighted average equation, 51
Weighted three point estimation, 50–51
Work breakdown structure (WBS)
cost management, 153–54
creating, 27–29
defined, 27
in defining activities, 37–38
example, 28, 154
Global (GWBS), 39
initial, 38
integrating to organization, 155–56
Working hours, 54
Work package planning, 38–39

Recent Titles in the Artech House
Effective Project Management Library

Robert K. Wysocki, Series Editor

The Project Management Communications Toolkit, Carl Pritchard

Project Management Process Improvement, Robert K. Wysocki

Critical Chain Project Management, Second Edition, Lawrence P. Leach

Practical Project Management for Engineers, Nehal Patel

For further information on these and other Artech House titles, including previously considered out-of-print books now available through our In-Print-Forever® (IPF®) program, contact:

Artech House
685 Canton Street
Norwood, MA 02062
Phone: 781-769-9750
Fax: 781-769-6334
e-mail: artech@artechhouse.com

Artech House
46 Gillingham Street
London SW1V 1AH UK
Phone: +44 (0)20 7596-8750
Fax: +44 (0)20 7630-0166
e-mail: artech-uk@artechhouse.com

Find us on the World Wide Web at: www.artechhouse.com